Ultra-high Frequency Linear Fiber Optic Systems

Kam Y. Lau

Ultra-high Frequency Linear Fiber Optic Systems

With 120 Figures

 Springer

Kam Y. Lau
Professor Emeritus
Department of Electrical Engineering and Computer Sciences
Cory Hall, University of California
Berkeley, CA 94720, USA

ISBN 978-3-540-25350-1   e-ISBN 978-3-540-49906-0

Library of Congress Control Number: 2008939870

Typesetting: Data prepared by the Author and by SPi
Cover design: WMX Design GmbH, Heidelberg

SPIN 10970423
Printed on acid-free paper

9 8 7 6 5 4 3 2 1

springer.com

To my late parents,

to Ching Elizabeth Ho and her parents

Foreword

It would seem that a comprehensive book on such an interesting and practically important topic as linear fiber-optic systems that includes an in-depth theoretical and practical treatment of their key enabling devices is long overdue. Linear or analog fiber-optic systems are important segments of optical transmission systems that makeup the global optical network. In addition, linear optical systems are important in military applications for sensing and for distributed antennas. A deep understanding and practical appreciation of the enabling technologies – both lasers and detectors – that support such systems is essential to fully appreciate and master the innovation and design of such systems. Professor Kam Lau's book provides an in-depth treatment of both linear fiber-optic systems and their key enabling devices.

The semiconductor laser is at the heart of such analog systems. As the engine in every optical communication system, the semiconductor laser is the light source that provides the single-, high-frequency carrier on which high bit-rate signals are impressed or encoded. While many optical transmission systems, including those that make up the long haul transoceanic and transcontinental, as well as metro and recent fiber-to-the-home networks, employ digital modulation of intensity "zeros" and "ones". Systems that use fiber as trunks to extend video signals deep into cable TV networks are analog systems. In such analog systems a video signal is directly impressed with high analog fidelity onto an optical carrier. For these analog systems high-speed modulation of lasers in which the modulated laser output is an undistorted (linear) replica of the video signal is critically important.

In addition to such analog application, in many digital systems – probably the majority from a numbers point of view – particularly at modest data rates or over shorter distance links, information encoding is achieved via direct on/off intensity modulation of the laser. Examples include cross-office links, data center links, storage area networks and fiber-to-the-home systems, to name a few. For these two large areas of applications, the high-speed modulation characteristics of semiconductor lasers are of fundamental importance.

Professor Kam Lau's book is focused on a fundamental understanding as well as the very practical implications of semiconductor laser performance in response to high-speed current modulation and on their application to high-frequency linear fiber-optic systems. The high-speed laser modulation information is also applicable to directly modulated digital optical transmission systems. The book draws substantially on the author's collaborative work with colleagues at Caltech and Ortel Corp. in the 1980s and on results generated in his research group in the 1990s as Professor of Electrical Engineering and Computer Sciences at University of California, Berkeley, which includes several seminal discoveries during these investigations. The author's prior industrial experiences as founding chief scientist of Ortel Corp. – a market leader in linear fiber-optic systems for the Hybrid-Fiber-Coax infrastructure (acquired by Lucent in 2000) and as a co-founder of LGC Wireless, Inc. – a market leader in in-building wireless coverage and capacity solutions (acquired by ADC Telecom in 2007) have clearly helped to provide a practical view at both the systems and device level which should be useful to the reader.

Parts I and II of the book are devoted to the Physics of High Speed Lasers. The author examines the properties of high-speed modulation, starting from first principles. Included are derivations for the frequency response as well as distortion effects important to analog system applications. The high-speed laser modulation performance is important for directly modulated digital fiber-optic systems as well.

Analog transmission systems are covered in Part III. Transmission impairment issues, including the impact of laser performance, are reviewed. Summaries of several experimental systems results, including those employing high frequency external modulators, provide a practical perspective. A particular example of employing analog links to provide indoor wireless signal distribution offers an interesting application that is of growing importance. The appendices nicely complement the main body of the book by including, e.g., background information on linear systems and alternate linear encoding approaches such as external waveguide modulators.

This book should serve as an excellent text for advanced graduate students engaged in research in high frequency fiber-optic links for cable TV and remote antenna systems as well as those interested in a fundamental understanding of high frequency laser modulation performance. It should also be valuable as a source reference for researchers and engineers in both academia and industry.

Holmdel, N.J., July 2008
Rod C. Alferness
Chief Scientist
Bell Laboratories, Alcatel-Lucent

Preface

Fiber-optics are firmly established as the dominant medium of terrestrial telecommunication infrastructure. Many excellent references and textbooks which treat this subject in great detail are available. Most of these books cover device and systems aspects of digital fiber-optic transmission. For this reason, digital fiber-optic systems will not be a subject of discussion at all in this book. In the current communication infrastructure, a sizeable portion of "access" traffic is carried by Hybrid-Fiber-Coax (HFC) infrastructure [1] which employs subcarrier transmission[1] (essentially an analog format)[2] to support both CATV[3] and cable modem Internet access. A similar situation exists in some military radar/communication systems where personnel and signal processing equipment is remoted from the physical antennas which are often in the way of homing weapons. The format of transmission in these systems

---

[1] Subcarrier transmission is essentially frequency division multiplex in the RF domain modulated on an optical carrier.

[2] Most subcarrier transmissions use QPSK or higher "-nary" modulation of the RF carrier, and are thus digital in content; but the criteria used to gauge the quality of the signals are still RF in nature. It is thus a matter of semantics or opinion whether subcarrier modulation is "analog" or "digital." The author prefers to interpret subcarrier modulation and transmission as "analog" because the principal criteria used to gauge their performances are analog in nature.

[3] "CATV" stands for Community Antenna TV, in which a large satellite antenna at a remote location with good reception of satellite signals, (typically in analog FDM format). The satellite antenna is often collocated with video processing and Internet access equipments. Collectively these facilities are known as the "Head End." Linear (analog) fiber optic links carry the signals to subdivision hubs. From there it is distributed to individual homes through a coaxial cable network – hence the name "Hybrid Fiber Coax." Employment of linear fiber-optic components and systems eliminates the need for serial placement of numerous high linearity RF amplifiers ("in-line amplifiers") to compensate for the high loss of coaxial cables in the long span from the head end to subdivision hubs. Failure of a single RF amplifier results in loss of service to an entire subdivision.

is also analog in nature. Various nomenclatures have been given to these systems, the most popular of which are – RF photonics, linear/analog lightwave transmission, the former being popular with the defense establishment, the latter with commercial establishments serving the HFC infrastructure.

While the most significant commercial application of linear fiber-optic systems has been the deployment of HFC infrastructure in the 1990s, the earliest field-installed RF fiber-optic transmission system was made operational in the late 1970s/early 1980s at the "Deep Space Network" (DSN) at Goldstone, Mojave Desert in Southern California, just north of Los Angeles. The DSN[4] is a cluster of more than a dozen large antenna dishes, the largest of which measures 70 m in diameter (Fig. P.1).

Fig. P.1. Aerial view of Deep Space Network (DSN) at Goldstone, Mojave Desert, Southern California. The giant 70 m dish is in the foreground, a dozen other smaller dishes are located around it spaced up to 10 km from one another

---

[4] The DSN antenna network consists of three clusters of large antennas each identical to the one at Goldstone, the other two are located near Madrid, Spain, and Canberra, Australia. These three DSN sites are located roughly evenly in longitude around the globe, to enable maximum round the clock coverage of the interplanetary space probes.

The DSN is operated by Jet Propulsion Laboratory (JPL) of Caltech and has been used by NASA (National Aeronautics and Space Administration) over the past two decades to track and communicate with unmanned space probes exploring the solar system to its very edge and beyond. In particular, the two "Voyager" space probes (Voyager I & II) were designed and destined to head out of the Solar system into interstellar space[5]. At >8 billion miles from earth, the signal power received by/from these interstellar spacecrafts is minuscule to the extent that even a single 70 m diameter giant antenna alone at DSN was inadequate to carry out the task of communication/tracking. A fiber-optic network was installed at the Goldstone DSN in the late 1970s/early 1980s, with the sole purpose of transmitting to all antennas in the network an ultra-stable microwave reference signal at 1.420405752 GHz (the 21 cm line of atomic hydrogen, accurate to within parts in 10^{15}, generated by a hydrogen maser in an environmentally controlled facility). All antennas in the network are synchronized to this ultra-stable frequency reference, enabling them to act as a single giant antenna using the phase array concept, capable of communicating with the spacecrafts as they head out to interstellar space.

This extreme stability requirement of microwave fiber-optic links necessitates active feedback stabilization techniques to compensate for any and all physical factors that can affect lightwave propagation in the optical fiber cable, including temperature, humidity and mechanical effects. A scheme to achieve this was disclosed in a patent, the front page of which is shown in Fig. P.2. Readers interested in the details of this stabilization scheme should consult the full patent, downloadable from the U.S. Patents and Trademark Office web site – http://www.uspto.gov/ and search for patent #4,287,606.

Among the numerous nomenclatures used to describe this type of analog fiber-optic transmission systems, "linear lightwave transmission" has gained traction over others, even though in substance there is no distinction between "linear lightwave transmission" and the more traditional, technically descriptive "analog/RF lightwave transmission," the rationale being that:

1. The financial community generally offers a higher reward to non-defense related businesses, presumably because the well-being of the latter depends too heavily on the often unpredictable international political climate
2. "Linear" lightwave systems surpass "analog" lightwave systems, in terms of marketability of hardware manufacturers to the financial community because the latter conjures up undesirable archaic impressions.

---

[5] On May 31, 2005 and August 30, 2007 Voyager I and Voyager II respectively passed the heliosphere, the critical boundary at 8.7 billion miles from the sun that marks the transition from the solar system into interstellar space. For more information, visit http://voyager.jpl.nasa.gov/.

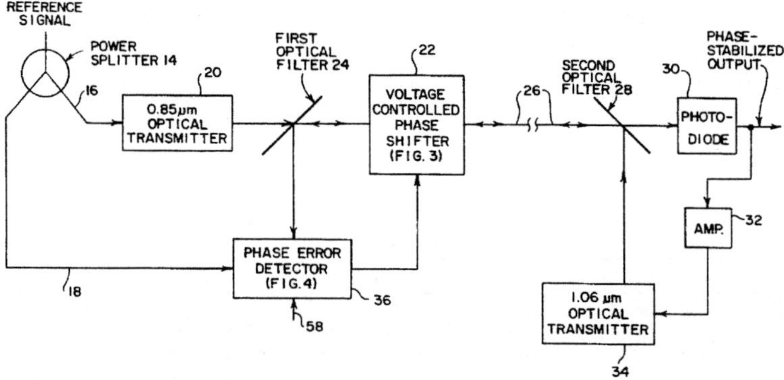

9/1/81   XR   4,287,606

United States Patent [19]

Lutes, Jr. et al.

[11]   **4,287,606**

[45]   **Sep. 1, 1981**

[54]   **FIBER OPTIC TRANSMISSION LINE STABILIZATION APPARATUS AND METHOD**

[76]   Inventors: **Robert A. Frosch,** Administrator of the National Aeronautics and Space Administration, with respect to an invention of **George F. Lutes, Jr.,** Glendale; **Kam Y. Lau,** Pasadena, both of Calif.

[21]   Appl. No.: **188,160**

[22]   Filed: **Sep. 17, 1980**

[51]   **Int. Cl.³** **H04B 9/00**
[52]   **U.S. Cl.** **455/617;** 455/610; 455/615; 455/612
[58]   **Field of Search** 455/610, 612, 615, 617

[56]   **References Cited**

U.S. PATENT DOCUMENTS

3,571,597	3/1971	Wood	455/607
3,887,876	6/1975	Zeidler	455/610
3,953,727	4/1976	d'Auria	455/610
4,102,572	7/1978	O'Meara	455/606
4,234,971	11/1980	Lutes	455/619

Primary Examiner—Howard Britton
Attorney, Agent, or Firm—Monte F. Mott; John R. Manning; Paul F. McCaul

[57]   **ABSTRACT**

A fiber optic transmission line stabilizer for providing a phase-stabilized signal at a receiving end of a fiber optic transmission line (26) with respect to a reference signal at a transmitting end of the fiber optic transmission line (26) so that the phase-stabilized signal will have a predetermined phase relationship with respect to the reference signal regardless of changes in the length or dispersion characteristics of the line (26). More particularly, a reference signal of RF frequency modulates a 0.85 micrometer wavelength optical transmitter (20). The output of the optical transmitter (20) passes through a first optical filter (24) and a voltage-controller phase shifter (22), the output of the phase shifter (22) being provided to the fiber optic transmission line (26). At the receiving end of the fiber optic transmission line (26), the signal is demodulated, the demodulated signal being utilized to modulate a 1.06 micrometer optical transmitter (34). The output signal from the 1.06 micrometer optical transmitter (34) is provided to the same fiber optic transmission line (26) and passes through the voltage-controlled phase shifter (22) to a phase error detector (36). The phase of the modulation of the 1.06 micrometer wavelength signal is compared to the phase of the reference signal by the phase error detector (36), the detector (36) providing a phase control signal related to the phase difference. This control signal is provided to the voltage controlled phase shifter (22) which alters the phase of both optical signals passing therethrough until a predetermined phase relationship between modulation on the 1.06 micrometer signal and the reference signal is obtained.

21 Claims, 8 Drawing Figures

Fig. P.2. Front page of patent disclosure detailing method and apparatus of fiber-optic transmission of ultra-stable frequency reference for antenna synchronization at NASA Deep Space Network

Cable TV distribution and associated cable modem Internet access really belong in the realm of "access" and not telecommunications, but they are nonetheless an important and integral part of present day communication infrastructure.

Yet another emerging means of access and enterprise private communication infrastructure construct is free-space point-to-point millimeter-wave ("mm-wave") links[6], capable of high data rates (in multi-Gb/s') due to the high carrier frequency in the mm-wave range. It also offers ease of construction – the only criteria being locating and gaining access to line-of-site vantage points for the transmitting and receiving antennae. The physical alignment of these mm-wave links are substantially more forgiving than corresponding free-space optical links, and in contrast to the latter, suffers only minimal free-space propagation impairment under less-than-ideal weather conditions. Another desirable factor for this type of system over a wire-line infrastructure, in addition to savings in construction cost and time, is avoidance of onerous issues of negotiating right-of-ways. This is all the more apparent, for example, in the situation of the construction of a high speed private data network between buildings within a corporate campus which spans across an interstate highway with no right-of-way access for private enterprise (Fig. P.3).

It is also desirable for this type of commercial systems to have the mm-wave transceiver equipment located remotely from the antenna sites, even though

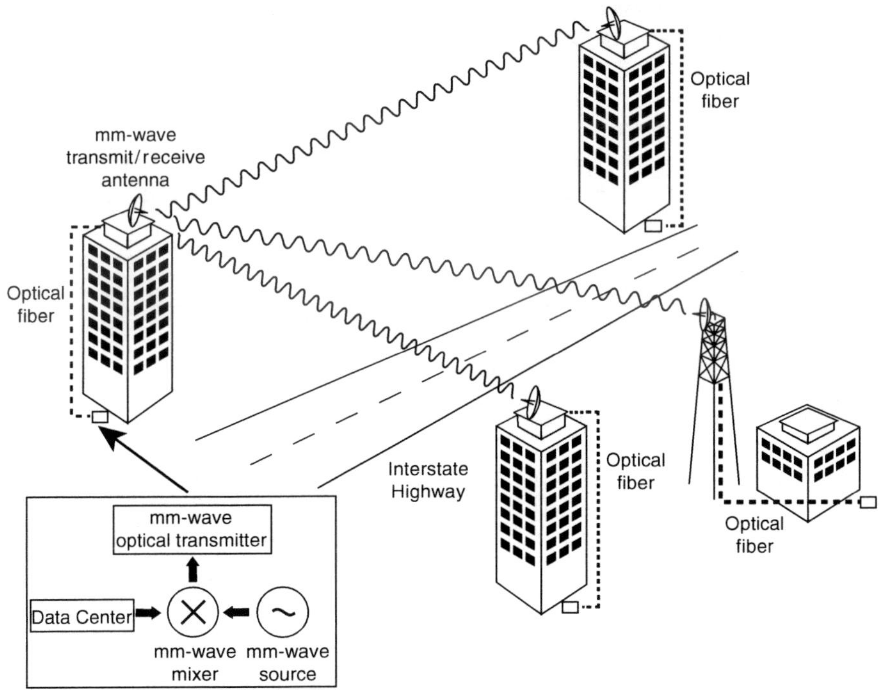

Fig. P.3. Schematic of a campus-scale free space mm-wave network

---

[6] Examples of commercial offerings of this type of products are –
http://www.loeacom.com/, retrieved: 2008/07/25, 4:25PM and
http://www.bridgewave.com/, retrieved: 2008/07/25, 4:26PM.

the antenna sites are not at risk of being targeted by homing destructive vehicles as in the case of military installations. The reasons are two-fold:

1. FCC stipulates that licensees of mm-wave bands hold their emission frequencies tightly, to the extent that economical free-running mm-wave oscillators operating in an uncontrolled outdoor environment do not suffice; they must be locked to a stable reference located remotely in an environmentally controlled location or else the data to be transmitted are "premixed" onto the mm-wave carrier at a remote location and then "piped" to the antenna sites for free space transmission;

2. The amount of service work required at the remote antenna sites (which are often difficult to access and subject to inclement weather conditions) can be minimized.

Given the above rationale, the problem then becomes that of innovating means of transporting mm-wave subcarrier signals over intermediate distances (up to ~10 s of km.) Use of mm-wave waveguides or coaxial cables are simply not viable due to dispersion/loss, in addition to being bulky, heavy and prohibitively expensive.

The use of optical fiber should be an ideal solution. A substantial portion of this book deals with the subject of modulating and transmitting mm-wave subcarrier signals on an optical carrier, and associated fiber transmission effects.

In traditional telecom infrastructure, long haul links use externally modulated laser transmitters to minimize optical frequency chirp and resultant signal degradation due to fiber dispersion. The "external" modulator in telecom long haul transmission is actually an electro-absorption (EA) modulator monolithically integrated with a 1.55 µm cw laser, the "IML" (Integrated Modulator Laser). Within the metropolitan arena the fiber infrastructure is dominated by 1.3 µm links employing directly modulated Distributed Feed-Back diode lasers, "DFBs." In terms of the raw number of laser transmitters deployed, directly modulated 1.3 µm lasers far outstrip that of "IML". The situation is similar in HFC networks, where directly modulated 1.3 µm linear laser transmitters transporting RF signals over longer spans within subdivisions are far more prevalent than long reach links feeding subdivision nodes from the "head-end". In terms of economics, directly modulated linear laser transmitters thus carry more weight. This is also the case with the fiber-optic distribution infrastructure supporting the type of mm-wave free-space interconnection networks as described above. For economic reasons, it is preferred that short spans for transporting mm-wave signals up to roof tops or tall towers/poles ("short-reach" links) employ 1.3 µm Single Mode Fibers "SMFs" using *inexpensive* directly modulated laser transmitters which are only *slight variants* of those deployed in telecom, where economic advantages of the latter's mass production capacity can be exploited, while "long-reach" links serving a wider region consist of 1.55 µm SMFs (*not* necessarily of the dispersion-shifted type.) Transmitters for these long-reach links employ cw laser diodes in conjunction with a high frequency external modulator, of which the velocity-matched Mach–Zehnder type is a logical choice. Similar to the case of HFC networks, the raw number of directly modulated 1.3 µm "short reach" links far outstrips that of "longer reach" 1.55 µm links requiring externally modulated transmitters. To reflect this practical reality, on top of the innovative challenges in constructing low-cost "telecom-type", directly modulated laser diodes capable of operating at mm-wave frequencies for the "short-reach" spans. A considerable portion of this book (Chaps. 8–11) is devoted to this subject.

"Longer reach" 1.55 µm links for transport of millimeter-wave signals up to 50–100 km employing external electro-optic Mach-Zehnder modulators are described in Chaps. 12 and 13, for dispersion-shifted and non-dispersion-shifted fibers respectively. The basic principle of high frequency velocity-matched modulators is presented in Appendix C.

The core material (Parts II and III) in this book represents research results on this subject generated by members of the author's research group in the 1990s in the Department of Electrical Engineering and Computer Sciences at U.C. Berkeley. These materials are augmented by discussions, in Part I of the book, of general baseband modulation of semiconductor lasers and associated fiber transmission effects which constitute the basis of understanding of directly modulated laser transmitter prevalent in metropolitan and local area fiber optic networks, as well as subcarrier fiber-optic links in HFC networks today.

These traditional baseband direct modulation approaches, however, do not appear to have the potential of extending into the mm-wave frequency range. Innovative approaches are therefore needed to accomplish the task of transporting mm-wave subcarrier modulated optical signals in optical fibers, as covered in Parts II and III of this book. Following a review of the current understanding of direct modulation of semiconductor lasers (Chaps. 1–6) and some related noise and impairments due to laser-fiber interaction (Chap. 7) in Part I of the book, Part II describes an innovative approach known as "resonant modulation", which is basically a "small-signal" version of the classic technique of active mode-locking applied in the mm-wave frequency range to *monolithic, standard telecom* laser diode structures for transmission of subcarrier signals at beyond the baseband limit (into the mm-wave frequency range). Part III discusses fiber transmission effects of mm-wave subcarrier signals in general (Chaps. 12, 13), as well as a high level systems perspective of a particular application to fiber-wireless coverage (Chap. 14). Chapter 15 discusses the effect and mechanism of superposition of a high frequency tone in the modulation current input to the laser for suppression of interferometric noises (such as modal noise in MMF (Multi-Mode Fiber) links or intensity noise generated by conversion from phase noise of the laser by multiple retroreflections in SMF (Single-Mode Fiber) links) by superposition of a high frequency tone in the modulation current input to the laser (Chap. 15). Part III concludes with another innovative, powerful approach to optical transmission of mm-wave subcarrier signals – "Feed-forward Modulation" (Chap. 16), which circumvents a significant disadvantage of the "resonant modulation" approach, namely that the laser device must be customized for transmission at a given mm-wave subcarrier frequency, and cannot be freely varied electronically thereafter. This is precisely the capability of "Feed-forward Modulation", albeit at the cost of higher complexity and part count.

Notes on common metrics of RF signal qualities can be found in Appendix A. Basic Principles of high speed photodiodes and narrow-band photoreceivers intended for subcarrier signals are discussed in Appendix B. Basic principles and state-of-the art external optical modulators are briefly reviewed in Appendix C.

Appendix D describes theoretical direct modulation response of "superluminescent lasers" – laser diodes with very low end mirror reflectivities operating at a very high internal optical gain, using the full nonlinear, spatially non-uniform traveling-wave rate equations, the computed response is compared to that of conventional laser diodes using the simple rate equations, the validity of the spatially uniform rate equations is thus established.

Acknowledgements

It was mentioned in the Preface that portions of the technical content of this book were generated by members of the author's research group during the 1990s in the Department of Electrical Engineering and Computer Sciences at U.C. Berkeley. They include (in alphabetical order) Drs. Lisa Buckman, Leonard Chen, David Cutrer, Michael Daneman, John Gamelin, John Georges, Janice Hudgings, Sidney Kan, Meng-Hsiung Kiang, Inho Kim, Jonathan Lin, Jocelyn Nee, John Park, Petar Pepeljugoski, Olav Solgaard, Dan Vassilovski, Bin Wu, and Ta-Chung Wu. Other parts of this book include work performed by the author in the 1970s and 1980s with various collaborators including Prof. Yasuhiko Arakawa, Drs. Nadav Bar-Chaim, Christoph Harder, Israel Ury, Prof. Kerry Vahala, and Prof. Amnon Yariv.

Special thanks go to Dr. John Park whose help in the editorial task was instrumental in bringing this book into being. Thanks are also due Kit Pang Szeto who undertook the massive task of typesetting this book in LaTeX from cover to cover, in addition to generating nearly all graphics and illustrations in this book. Others who have contributed to the typesetting and graphics work include Wai See Cheng and Tsz Him Pang.

The author expresses his thanks to each and everyone mentioned above, without whom this book could not have possibly materialized.

Berkeley, CA, June 2008   *Kam Y. Lau*

Contents

Part I

Physics of High Speed Lasers

1

Basic Description of Laser Diode Dynamics by Spatially Averaged Rate Equations: Conditions of Validity

A laser diode is a device in which an electric current input is converted to an output of photons. The time dependent relation between the input electric current and the output photons are commonly described by a pair of equations describing the time evolution of photon and carrier densities inside the laser medium. This pair of equations, known as the laser rate equations, will be used extensively in the following chapters. It is therefore appropriate, in this first chapter, to summarize the results of Moreno [2] regarding the conditions under which the rate equations are applicable.

1.1 The "Local" Rate Equations

The starting point for the analysis of laser kinetics involves the coupled rate equations which are basically *local* photon and injected carrier conservation equations [3]:

$$\frac{\partial X^+}{\partial t} + c\frac{\partial X^+}{\partial z} = A(N - N_{\mathrm{tr}})X^+ + \beta\frac{N}{\tau_{\mathrm{s}}}, \tag{1.1a}$$

$$\frac{\partial X^-}{\partial t} - c\frac{\partial X^-}{\partial z} = A(N - N_{\mathrm{tr}})X^- + \beta\frac{N}{\tau_{\mathrm{s}}}, \tag{1.1b}$$

$$\frac{\mathrm{d}n}{\mathrm{d}t} = \frac{J}{ed} - \frac{N}{\tau_{\mathrm{s}}} - A(N - N_{\mathrm{tr}})(X^+ + X^-), \tag{1.1c}$$

where z is the spatial dimension along the length of the laser, with reflectors of (power) reflectivities R placed at $z = \pm L/2$, X^+ and X^- are the forward and backward propagating photon densities (which are proportional to the light intensities), N is the local carrier density, N_{tr} is the electron density where the semiconductor medium becomes transparent, c is the group velocity of the waveguide mode, A is the gain constant in $\mathrm{s}^{-1}/(\text{unit carrier density})$, β is the fraction of spontaneous emission entering the lasing mode, τ_{s} is the spontaneous recombination lifetime of the carriers, z is the distance along the active medium with $z = 0$ at the center of the laser, J is the pump current density, e is the electronic charge and d the thickness of the active region in which the carriers are confined. For the remaining of this chapter it is assumed that $N_{\text{tr}} = 0$, the only consequence of which is a DC shift in the electron density, which is of significance only in considering lasing threshold. In addition, the following simplifying assumptions are made in writing down (1.1):

1. The quantities X^* describe the local photon number densities of a longitudinal mode of the passive laser cavity at a given (longitudinal) position in the laser cavity, at time t, integrated over the lasing linewidth of the longitudinal mode, which is assumed to be much narrower than the homogeneously broadened laser gain spectrum.
2. The gain coefficient (AN) is a linear function of the injected carrier density N (A is popularly known as the "differential optical gain coefficient" and will be shown in later chapters to play a key role in determining direct modulation bandwidth of laser diodes).
3. Variations of the carrier and photon densities in the lateral dimensions are neglected.
4. Diffusion of carriers is ignored.

Assumptions 1 and 2 are very reasonable assumptions which can be derived from detailed analysis [4–6]. The representation of the semiconductor laser as a homogeneously broadened system can also be derived from basic considerations [7]. Transverse modal and carrier diffusion effects, ignored in assumptions 3 and 4, can lead to modifications of the dynamic behavior of lasers [8,9].

Equations (1.1) are to be solved subject to the boundary conditions

$$X^- \left(\frac{L}{2}\right) = RX^+ \left(\frac{L}{2}\right), \tag{1.2a}$$

$$X^+ \left(\frac{-L}{2}\right) = RX^- \left(\frac{-L}{2}\right). \tag{1.2b}$$

The steady state solution of (1.1) gives the static photon and electron distributions inside the laser medium, and has been solved analytically in [4]. The solution is summarized as follows, where the zero subscript denotes steady state quantities:

$$X_0^+(z) = \frac{ae^{u(z)} - \beta}{Ac\tau_{\text{s}}}, \tag{1.3a}$$

$$X_0^-(z) = \frac{ae^{-u(z)} - \beta}{Ac\tau_{\text{s}}}, \tag{1.3b}$$

where a is a quantity given by the following transcendental equation:

$$(1 - 2\beta)\xi + 2a \sinh \xi = \frac{gL}{2}, \tag{1.4}$$

where

$$\xi = \frac{1}{2}\sqrt{\frac{(R-1)^2\beta^2}{(Ra)^2} + \frac{4}{R}} + (R-1)\frac{\beta}{Ra} \tag{1.5}$$

and $g = AJ_0\frac{\tau_s}{ed}$ is the unsaturated gain, and $u(z)$ is given transcendentally by

$$(1 - 2\beta)u(z) + 2a\sinh u(z) = gz. \tag{1.6}$$

The electron density $N_0(z)$ is given by

$$AcN_0(z) = \frac{g}{1 + 2a\cosh u(z) - 2\beta}. \tag{1.7}$$

Figure 1.1 shows plots of $X_0^+(z)$, $X_0^-(z)$ and $g_0(z) = AcN_0(z)$ for a $300\,\mu\text{m}$ laser with three values of end mirror reflectivities, (a) 0.3, (b) 0.1 and (c) 0.9. The high non-uniformity in the distributions becomes apparent at low reflectivities.

1.2 Spatially Averaged Rate Equations and Their Range of Validity

Equations (1.1) constitute a set of three coupled non-linear differential equations in two variables which do not lend themselves to easy solutions. Considerable simplification can be made if the longitudinal spatial variable (z) is integrated over the length of the laser. Such simplification is valid only when the end mirror reflectivity is "sufficiently large", A more precise definition of the range of validity of such an assumption is given in the following, summarizing the approach of [2].

To begin, (1.1a) and (1.1b) are integrated in the z variable, resulting in

$$\frac{\mathrm{d}x^{+*}}{\mathrm{d}t} + c\left[X^+\left(\frac{L}{2}\right) - X^+\left(\frac{-L}{2}\right)\right] = A(NX^+)^* + \beta\frac{N^*}{\tau_s}, \tag{1.8a}$$

$$\frac{\mathrm{d}x^{-*}}{\mathrm{d}t} - c\left[X^-\left(\frac{L}{2}\right) - X^-\left(\frac{-L}{2}\right)\right] = A(NX^-)^* + \beta\frac{N^*}{\tau_s}, \tag{1.8b}$$

where * denotes the spatial average $\int_{-L/2}^{L/2}\frac{\mathrm{d}z}{L}$. Adding (1.8a) and (1.8b),

$$\frac{\mathrm{d}p^*}{\mathrm{d}t} + \frac{2c(1-R)P(L/2)}{L(1+R)} = A(NP)^* + 2\beta\frac{N^*}{\tau_s}, \tag{1.9}$$

where $P = X^+ + X^-$ is the total local photon density and the boundary conditions (1.2) have been used. Equation (1.1c) integrates straightforwardly to

$$\frac{\mathrm{d}n^*}{\mathrm{d}t} = \frac{J}{ed} - \frac{N^*}{\tau_s} - A(NP)^*, \tag{1.10}$$

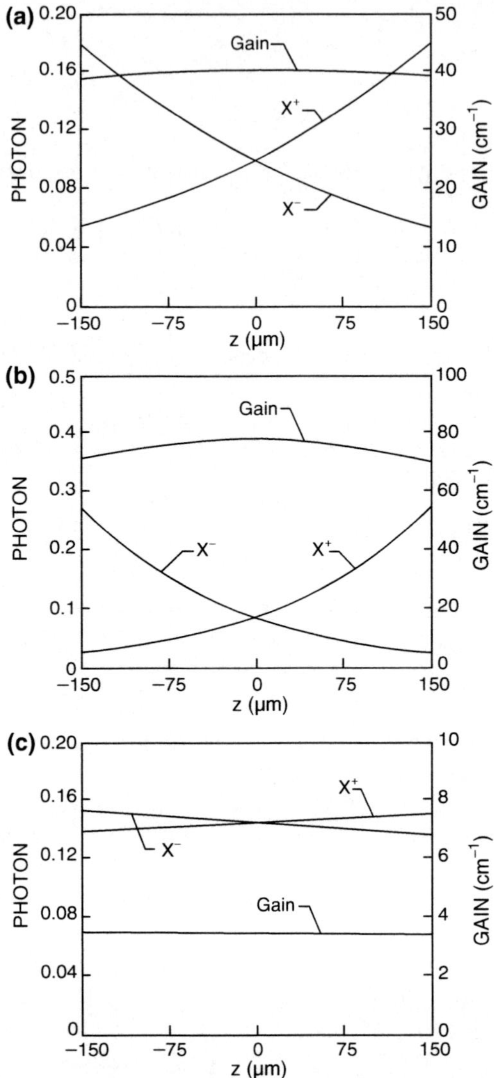

Fig. 1.1. Steady state photon and electron density distributions inside laser diodes with mirror reflectivities of (**a**) 0.3, (**b**) 0.1, and (**c**) 0.9

where a uniform pump current of density J is assumed. A is known as the "differential optical gain". It will be shown in later chapters to play a key role in determining direct modulation bandwidth of laser diodes. Introducing factors f_1 and f_2 as follows:

$$f_1 = \frac{(NP)^*}{N^* P^*},$$

(1.11)

$$f_2 = \frac{P(L/2)}{P^*(1+R)}, \tag{1.12}$$

one can write the spatially averaged rate equations (1.9) and (1.10) in the following form:

$$\frac{dp^*}{dt} = Af_1 N^* P^* - 2c(1-R)f_2 \frac{P^*}{L} + 2\beta \frac{N^*}{\tau_s}, \tag{1.13}$$

$$\frac{dn^*}{dt} = \frac{J}{ed} - \frac{N^*}{\tau_s} - Af_1 N^* P^*, \tag{1.14}$$

which are recognized as the commonly used rate equations [10, 11] if the conditions

$$f_1 = 1, \tag{1.15}$$

$$f_2 = -\frac{1}{2}\frac{\ln R}{1-R}, \tag{1.16}$$

are satisfied. The first of these conditions requires, for the quantities N and P, that the spatial average of the product equals the product of the spatial averages. This condition is not satisfied in general, but it will be true if the electron density N is uniform, as in the case when R approaches unity, which is apparent from Fig. 1.1c. The second condition requires the photon loss rate in (1.13) to be inversely proportional to the conventional photon lifetime. It will also be satisfied if R is very close to unity, since both (1.12) and (1.16) converge to 0.5 this limit.

A more precise delineation of the range of the applicability of conditions (1.15) and (1.16) is obtained by calculating f_1 and f_2 from exact steady state solutions (1.3) through (1.7), and comparing them with (1.15) and (1.16). From (1.3) and (1.7),

$$f_1 = \frac{L \int \frac{P}{1+A\tau_s P} dz}{\int \frac{dz}{1+A\tau_s P} \int P dz}, \tag{1.17}$$

$$f_2 = \frac{LX^+(L/2)}{\int P dz}, \tag{1.18}$$

where the integrals are evaluated over the length of the laser. These integrals can be numerically evaluated using (1.3) through (1.7), and the results are shown in Fig. 1.2. Figure 1.2 shows numerically computed plots (solid lines) of f_1 and $1/f_2$ as a function of end mirror reflectivity R, the calculation was done with the laser biased above threshold. The dotted lines are the "ideal" values of f_1 and f_2 given by (1.15) and (1.16). The figure indicates that the usual rate equations are reasonably accurate for R larger than approximately 0.2 – valid for laser diodes constructed from III–V materials, which have facet reflectivities of $\approx$0.3.

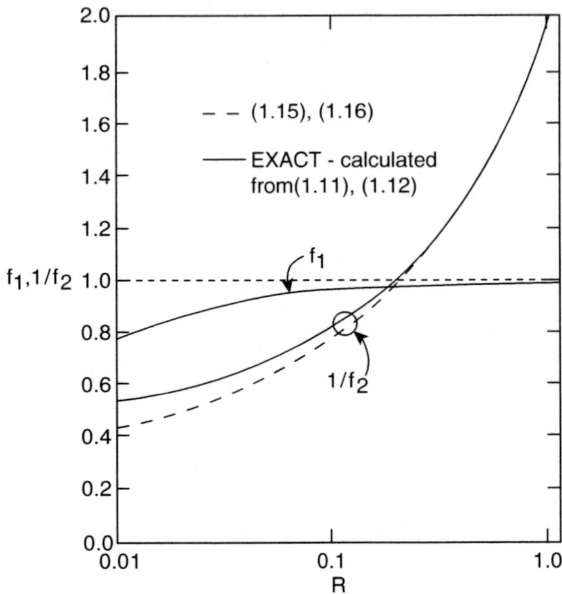

Fig. 1.2. Variations of f_1 and $\frac{1}{f_2}$ with R when $\beta \leq 10^{-3}$ and $gL > 10$

The above results lead to the conclusion that the simple rate equations, expressed in (1.19) and (1.20) (where the N and P now denote *averaged* quantities, in the *longitudinal spatial dimension*):

$$\frac{\mathrm{d}n}{\mathrm{d}t} = \frac{J}{ed} - \frac{N}{\tau_{\mathrm{s}}} - ANP \tag{1.19}$$

$$\frac{\mathrm{d}p}{\mathrm{d}t} = ANP - \frac{P}{\tau_{\mathrm{p}}} + \beta\frac{N}{\tau_{\mathrm{s}}} \tag{1.20}$$

$(1/\tau_{\mathrm{p}} = c/(2L)\ln(1/R)$ is the classical photon lifetime and $A = \kappa c)$ are reasonable representations if the end mirror reflectivity is above 0.2 and the laser is above threshold. The spontaneous emission factor β in (1.20) is a factor of two higher than that defined in (1.1) due to the inclusion of photons propagating in both directions. Common GaAs or quaternary lasers, with the mirrors formed by the cleaved crystal facets, have a reflectivity of ~0.3 and are thus well within the scope of (1.19) and (1.20). In Appendix D, the exact small signal version of (1.1) will be solved numerically and it will be found that (1.19) and (1.20) can very accurately describe the small signal frequency response of the laser for end mirror reflectivities as low as 10^{-3}. This is certainly not expected from a physical standpoint and serves as a surprise bonus for this simplification.

Another factor that can render the spatially-uniform assumption invalid is when "fast" phenomena, occurring on the time scale of a cavity transit time, are being considered. It is obvious that the concept of "cavity lifetime" and that of cavity modes, appearing in (1.20), are no longer applicable on that time scale. In common semiconductor lasers where the cavity length is approximately $300\,\mu$m. the cavity transit time is about 3.5 ps. The usual rate equations are therefore not applicable in describing phenomena shorter than about 5 ps, or at modulation frequencies higher than 60 GHz. Modulation regimes in the millimeter wave frequencies can take advantage of this cavity round-trip effect and is known as "resonant modulation", to be discussed in detail in Part II of this book.

In the following chapters, (1.19) and (1.20) will be used heavily, and serve as the basis for most of the analysis of the direct modulation characteristic of lasers.

2

Basic "Small Signal" Modulation Response

Most predictions of direct modulation response behavior of laser diodes are derived from a small signal analysis of the spatially averaged rate equations (1.19) and (1.20). This approach involves the assumption that the laser diode is driven by a "small" sinusoidal current at frequency ω, superimposed on a DC bias current: $J(t) = J_0 + j(\omega)\exp(i\omega t)$. The photon and electron density variables, n and p, are assumed to similarly consist of a "steady state" part, and a "small" time-varying part: $n(t) = n_0 + n(t)$; $p(t) = p_0 + p(t)$. Furthermore, the "small" time-dependent part is assumed to be sinusoidally varying in time, at the same frequency as the modulating current, i.e., $n(t) = n(\omega)\exp(i\omega t)$, $p(t) = p(\omega)\exp(i\omega t)$, where $n(\omega)$, $p(\omega)$ are complex quantities in general, thus incorporating the relative phase shifts between the drive current and the electron and phot on responses. As for what constitutes a "small" signal, one examines (1.19) and (1.20), and observes that the difficulty which prevents a simple analytic solution originates from the product term "np", present in both equations. A well known mathematical technique for obtaining an approximate solution is to first solve the equations in the "steady state", assuming no time variations in $J(t)$; i.e., $j(\omega) = 0$ consequently there would be no time variations in n and p either – $n(\omega) = p(\omega) = 0$. One then simply solves for n_0 and p_0 as a function of J_0. It turns out that the solutions thus obtained are incredibly simple, namely, $n_0 = 1$ and $p_0 = J_0 - 1$ if $J_0 > 1$ whereas, $n_0 = J$ and $p_0 = 0$ if $J_0 < 1$. A straight forward physical interpretation of these simple results is that the quantity $J_0 = 1$ represents the *lasing threshold current* of the laser. Thus, in the steady state when no modulation current is applied, the relation between the optical output power from the laser (which is proportional to P_0) to the input current J_0, is simply: $P_0 = 0$ if $J_0 < 1$; $P_0 = J_0 - 1$ if $J_0 > 1$.[1] There is a "knee" in the optical output

---

[1] These simple results have assumed the following normalization of the parameters – N is normalized by $(1/A\tau_{\mathrm{p}})$; P by $(1/A\tau_{\mathrm{s}})$; t by τ_{s} and J by $ed/(A\tau_{\mathrm{s}}\tau_{\mathrm{p}})$, in addition to ignoring the fact that the electron density must reach a certain value before the laser medium experiences positive optical gain – which is the assumption that $N_{\mathrm{tr}} = 0$, as mentioned in Chap. 1, in the derivation of (1.19) and (1.20). It is a simple matter to add a constant to the result above for the steady state value of the electron density to account for this fact.

power vs. input current relationship. Above this knee the output optical power is a *strictly linear* function of input current – in principle, assuming absence of any device imperfections. Thus from the above simple considerations the direct modulation characteristic of the laser should be strictly linear and free of distortions.

It turns out that this conclusion is only valid for modulation at "low" frequencies, which is to be expected since the conclusion is drawn from a steady state solution of the rate equations. Chapter 3 will examine the full frequency dependence of various modulation distortions. It will be shown in Chap. 3 that, apart from those induced by device imperfections. The product term $n(t)p(t)$ in the rate equations (1.19) and (1.20), due to *fundamental stimulated emission* responsible for the laser action, is the fundamental source of nonlinear distortions in directly modulated laser diodes. This thus establishes a *fundamental lower limit* to the amount of distortion generated in direct modulation of the laser diode which *cannot* be removed by means of clever device design. Consequently all ultra-linear fiber-optic transmitter which employs directly modulated laser diode must use some form of electrical distort ion-compensation techniques to correct for the laser modulation distortion.

The "small signal analysis" procedure then involves substituting $J(t)$, $n(t)$ and $p(t)$ as assumed above into the rate equations (1.19) and (1.20), followed by discarding products of the "small" terms, $n(\omega)p(\omega)$. It is for this reason that the "small signal" analysis is synonymous with a "linearization" analytic procedure.

For most operations of laser diodes the laser diode is "biased" with a DC current *above* lasing threshold. An exact numerical solution of the coupled rate equations (1.19) and (1.20) shows that severe "ringing" in the optical output $(p(t))$, occurs when the laser is turned on from below threshold. While this *may* be compensated for in *some* digital transmission links by electrical filtering of the signal at the receiver, this is totally unacceptable for linear transmission systems. It is thus assumed, in practically all discussions in the following chapters that the laser is "biased" with a DC current well above lasing threshold, a "small" modulation current is then superimposed on the DC bias current.

For a drive current which takes on a sinusoidal form at a certain frequency it is assumed that the condition "small-signal" is satisfied when *both* the electron and photon densities follow an exact sinusoidal variation at the same frequency. The issue of distortion in the modulated output photon density, which is a matter of prime importance in linear (analog) transmission, such as multi-channel CATV will be discussed in Chap. 3. Distortions which can be treated as (small) perturbations from the ideal sinusoidal responses fall under the "small signal" regime, as discussed in Chap. 3. Gross departures from the ideal sinusoidal response must be treated differently, often numerical procedures are required, as is often the case in digital on/off modulation.

Using this "small signal" approach, one reduces the coupled nonlinear rate equations (1.19) and (1.20) to two coupled *linear* differential equations, which are then further simplified by canceling out the (common) harmonic dependence of the variables, thus leaving two coupled linear *algebraic* equations in the variables $n(\omega)$, $p(\omega)$ with the drive term being $j(\omega)$. The commonly defined "frequency response" of the laser is $f(\omega) = p(\omega)/j(\omega)$, obtained easily by solving the two coupled *linearized (now algebraic)* rate equations. The form of $f(\omega)$ is $f(\omega) \sim 1/[(\mathrm{i}\omega)^2 + \gamma(\mathrm{i}\omega) - (\omega_\mathrm{R})^2]$, which is the classic form of a conjugate pole-pair second order low pass filter, exhibiting a flat low frequency pass band followed by a resonance peak at $f = f_\mathrm{R} = \omega_\mathrm{R}/(2\pi)$ before falling off at a rate of $-40\,\mathrm{dB/dec}$ (Fig. 2.1). The resonance peak at $\omega = \omega_\mathrm{R}$ is commonly known as the "relaxation oscillation" of the laser. This "relaxation oscillation" resonance peak in the frequency response can be interpreted as the frequency domain manifestation of the time-domain "ringing" of the optical output when the laser is driven from below threshold, as described before. When biased above threshold and modulated in the "small signal" regime the useful modulation bandwidth of semiconductor lasers is widely accepted to be equal to f_R, the relaxation oscillation frequency, although it is well recognized that the relaxation resonance, the magnitude of which varies

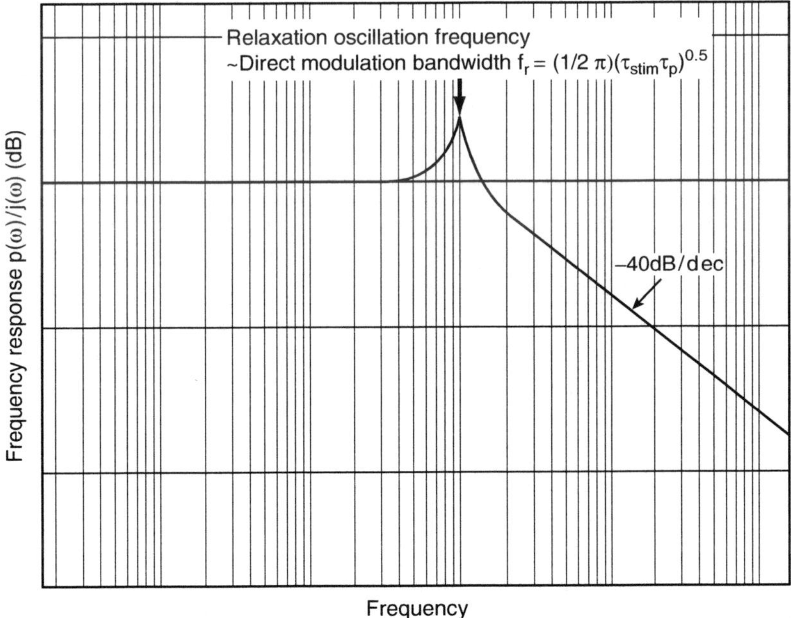

Fig. 2.1. Ideal "small-signal" modulation frequency response of a semiconductor laser (second-order low-pass filter function)

considerably among lasers [12, 13], can limit the useful bandwidth to somewhat below f_R. Nevertheless, as a standard for comparison, the modulation
bandwidth is simply taken as f_R. The relaxation oscillation frequency can be
obtained by a standard small-signal analysis of (1.19) and (1.20) (with the
only approximation being $\beta \ll 1$) which gives

$$\nu_{rel} = \frac{1}{2\pi}\sqrt{\frac{Ap_0}{\tau_p}}, \tag{2.1}$$

where p_0 is the steady state photon density in the active region.

Additional insight can be gained by noting that (2.1) can be rewritten as
$\nu_{rel} = (1/2\pi)(\tau_{stim}\tau_p)^{-1/2}$, i.e., the relaxation oscillation frequency is equal
to the inverse of the *geometric mean* of the photon and *stimulated* carrier
lifetimes.

Equation (2.1) suggests three obvious *independent* ways to increase the
relaxation frequency – by increasing the differential optical gain coefficient (A)
or the photon density, or by decreasing the photon lifetime. The differential
gain coefficient (A) can be increased roughly by a factor of 5 by cooling the
laser from room temperature to $77\,\text{K}$ [14] – even though this approach is
hardly feasible in practice, it can be used as a convenient means to verify the
validity of (2.1), as shown in Sect. 5.1.1. Biasing the laser at higher currents
would increase the photon density in the active region, which simultaneously
increases the optical output power I_{out} according to

$$I_{out} = \frac{1}{2}p_0\hbar\omega \ln\frac{1}{R}. \tag{2.2}$$

Short wavelength lasers (GaAs/GaAlAs lasers) used in LAN data links can
suffer catastrophic damages of the mirror facet at about $1\,\text{MW}\,\text{cm}^{-2}$. But long
wavelength quaternary lasers used for metropolitan networks or for telecom
do not suffer catastrophic mirror damages; but they do suffer from thermal-
related effects which reduce the modulation efficiency of the laser output, in
addition to a reduction in differential optical gain, in turn leading to a reduced
modulation bandwidth.

The third way to increase the modulation bandwidth is to reduce the photon lifetime by decreasing the length of the laser cavity. Such a laser needs
to be driven at higher current densities and thermal effects due to excessive heating will limit the maximum attainable modulation bandwidth. To
illustrate these tradeoffs the relaxation frequency as a function of the cavity
length and pump current density is plotted in Fig. 2.2a using (2.1) together
with the static solutions of (1.19) and (1.20). Also plotted in Fig. 2.2 is the
power density at the mirror using (2.2). As an example, a common GaAs laser
with a cavity of length $300\,\mu\text{m}$ operating at an output optical power density

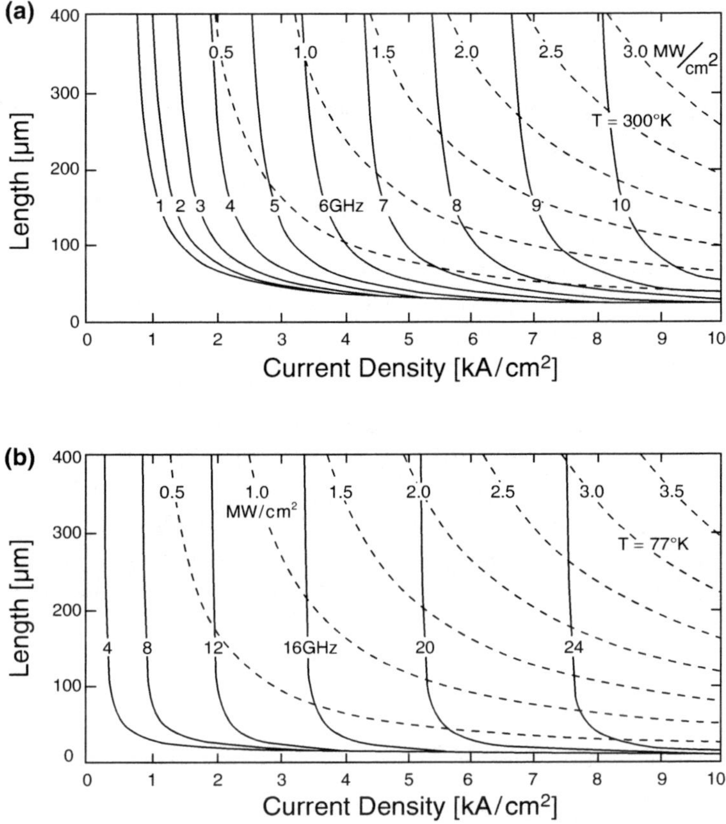

Fig. 2.2. (a) Relaxation frequency ν_{rel} (*solid lines*) and optical power density outside the mirrors (*dashed lines*) as a function of the cavity length and pump current density at $T = 300\,\mathrm{K}$. The following parameters are used: active layer thickness $= 0.15\,\mu\mathrm{m}$, $\alpha = 40\,\mathrm{cm}^{-1}$, $R = 0.3$, $c = 3 \times 10^9\,\mathrm{cm\,s}^{-1}$, $A = 2.56 \times 10^{-6}\,\mathrm{cm}^3\,\mathrm{s}^{-1}$, $\Gamma = 0.5$, $N_{\mathrm{om}} = 1 \times 10^{18}\,\mathrm{cm}^{-3}$, $B = 1.5 \times 10^{-10}\,\mathrm{cm}^3\,\mathrm{s}^{-1}$, $\hbar\omega = 1.5\,\mathrm{eV}$. (b) same as (a) but at $T = 77\,\mathrm{K}$. The same parameters as in (a) are used except: $A = 1.45 \times 10^{-5}\,\mathrm{cm}^3\,\mathrm{s}^{-1}$, $N_{\mathrm{om}} = 0.6 \times 10^{17}\,\mathrm{cm}^{-3}$, $B = 11 \times 10^{-10}\,\mathrm{cm}^3\,\mathrm{s}^{-1}$ [14]. (From [15], ©1983 AIP. Reprinted with permission)

of $0.8\,\mathrm{MW\,cm}^{-2}$ (close to catastrophic mirror damage, unless special provisions are taken) possesses a bandwidth of $5.5\,\mathrm{GHz}$, and the corresponding pump current density is $3\,\mathrm{kA\,cm}^{-2}$. Operating at an identical power density, the bandwidth is $8\,\mathrm{GHz}$ for a shorter laser with a cavity length of $100\,\mu\mathrm{m}$, but the corresponding current density is $6\,\mathrm{kA\,cm}^{-2}$. A higher current density alone may not be a cause for rapid degradation of lasers. For example, lasers with increased optical damage threshold as described above can operate at increased current densities without appreciable degradation of their reliability.

Figure 2.2b shows similar plots as in Fig. 2.2a but for a laser operating at liquid nitrogen temperature. The increase in bandwidth is a direct result of the increase in A. It can be seen that a modulation bandwidth beyond 20 GHz can be achieved, however, use of a short optical cavity and/or incorporation of a "non-absorbing window" structure is imperative under these operating conditions.

Experiments have been performed to determine the modulation bandwidth attainable in a short-cavity laser. The lasers used were buried heterostructure lasers fabricated on a semi-insulation substrate "BH on SI" [16]. In addition to a low lasing threshold (typically ≤ 15 mA) which is necessary to avoid excessive heating when operated at high above threshold, these lasers posses very low parasitic capacitance [17] which otherwise would obscure modulation effects at higher frequencies (>5 GHz). The lasers were mounted on a 50 Ω stripline. Microwave s-parameter measurements show that electrical reflection from the laser diode accounts to no more than a few dB (<5 dB) of variation in the drive current amplitude over a frequency range of 0.1–8.5 GHz. A sweep oscillator (HP8350) was used in conjunction with a network analyzer (HP8746B) to obtain the modulation data. The photodiode used was a high-speed GaAs *pin* diode fabricated on semi-insulating substrate. Its response was carefully calibrated from 0.1 to 10 GHz using a step-recovery-diode excited GaAs laser, which produced optical pulses 25 ps in full width at half-maximum, as measured by standard nonlinear autocorrelation techniques. The response of the diode to the optical pulse, recorded on the microwave spectrum analyzer, is then de-convolved by the finite width of the optical pulse. The observed modulation response of the laser is normalized by the photodiode response at each frequency. Figure 2.3a,b shows the cw light vs. current characteristic of a short-cavity (120 μm) BH on SI laser, and the modulation responses at various bias points as indicated in Fig. 2.3a are shown in Fig. 2.3b. The modulation bandwidth can be pushed to beyond 8 GHz as the bias point approaches the catastrophic damage level. Figure 2.4 shows the relaxation oscillation frequency of this laser as a function of $\sqrt{P_0}$, where P_0 is the output power, together with that of similar lasers with longer cavity lengths. All lasers tested suffered catastrophic damage between 6 and 8 mW per facet. The advantage of a short-cavity laser in high-frequency modulation is evident.

It is clear, from the above theoretical and experimental results, that, at least for short wavelength GaAs/GaAlAs lasers an ideal high-frequency laser should be one having a short cavity with a window structure, and preferably operating at low temperatures. This would shorten the photon lifetime, increase the intrinsic optical gain and the internal photon density without inflicting mirror damage. An absolute modulation bandwidth (at the point of catastrophic failure) of >8 GHz has already been observed in a 120 μm laser without any special protective window structure at room temperature. For reliable operation, however, the laser should be operated at only a fraction of its catastrophic failure power. That fraction depends on the specific laser

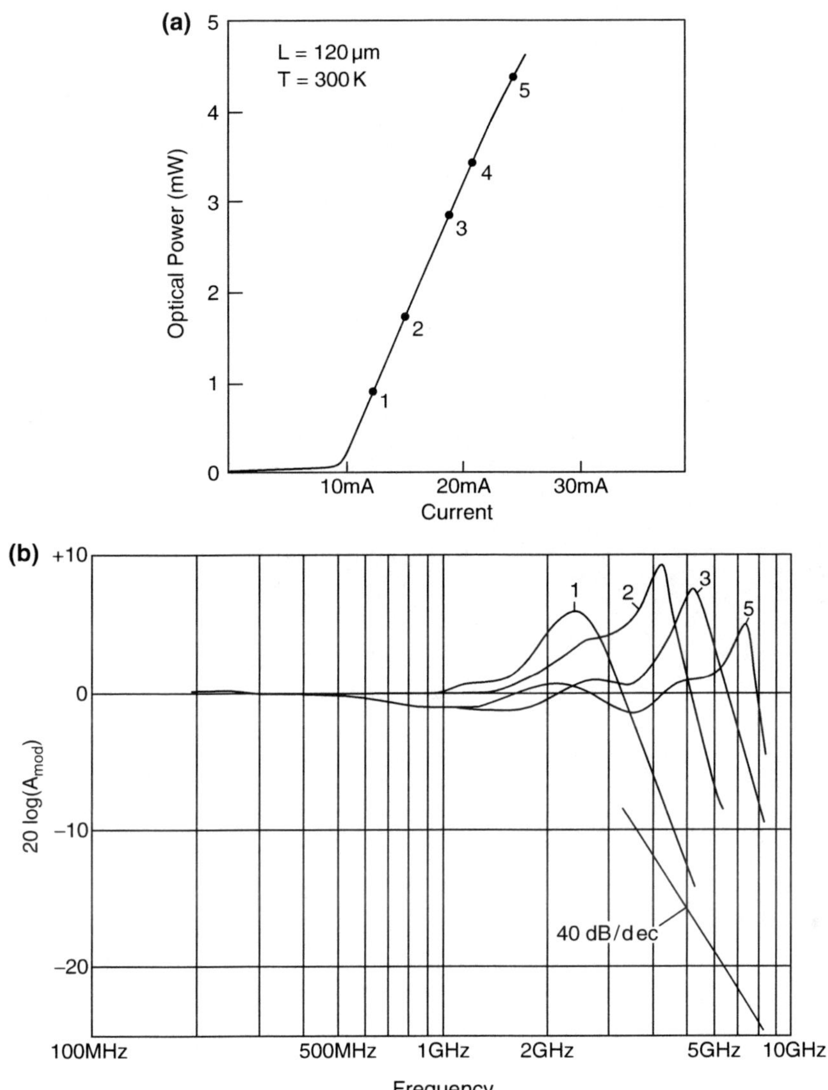

Fig. 2.3. (a) cw light vs. current characteristics of a BH on SI laser. Length of laser $= 120\,\mu$m. Modulation characteristics of this laser at various bias points indicated in the plot are shown in (b). (From [15], ©1983 AIP. Reprinted with permission)

structure and amounts to 1/2 to 1/3 for commercial devices of comparable construction [18]. This would place the useful modulation bandwidth of these short-cavity BH on SI lasers between 4.6 and 5.7 GHz. The same laser at 77 K without a window should have a modulation bandwidth of $\approx$12 GHz.

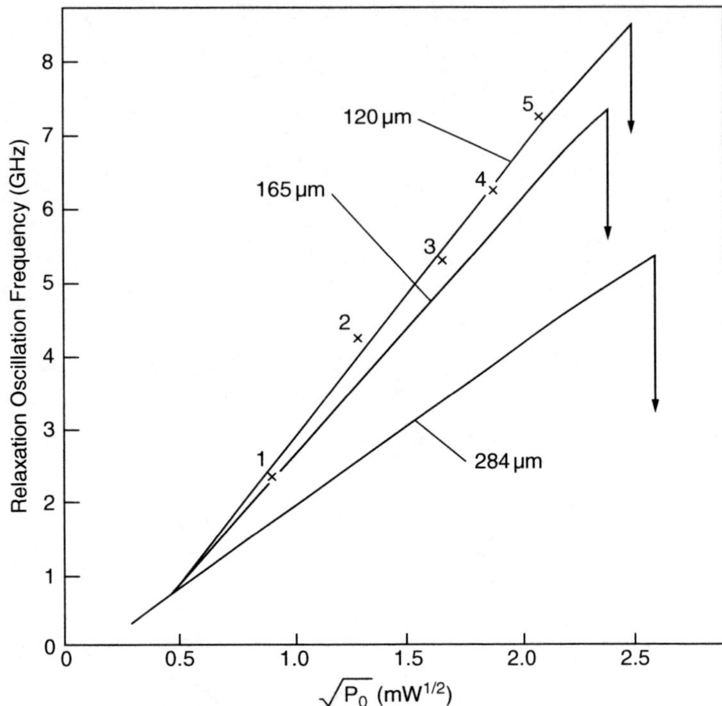

Fig. 2.4. Measured relaxation oscillation resonance frequency of lasers of various cavity lengths, as a function of $\sqrt{P_0}$ where P_0 is the cw output optical power. The points of a catastrophic damage are indicated by *downward pointing arrows*. (From [15], ©1983 AIP. Reprinted with permission)

3

Distortions in Direct Modulation
of Laser Diodes

3.1 Perturbation Analytic Prediction of Distortions
in Directly Modulated Laser Diodes

For analog transmission systems, linearity is a prime parameter. In fiber optics systems, the general modulation responses of the laser diode are well known [19] and their harmonic distortion characteristics have been theoretically analyzed [20–22]. This chapter describes a "perturbation" analytic approach for obtaining closed form solutions of harmonic distortions generated in the modulated optical intensity output of a laser diode under a pure single-tone sinusoidal modulation. The predictions are corroborated with experimental studies.

The approach described here is based on that described in [21, 23] and uses Fourier series expansion to solve the nonlinear laser rate equations for the photon density s and electron density n.

For the purpose of this analysis absolute time scale is not important and it is convenient to normalize the variables in (1.19) and (1.20) as follow – N is normalized by $(1/A\tau_{\mathrm{p}})$; P by $(1/A\tau_{\mathrm{s}})$; t by τ_{s} and J by $(ed/A\tau_{\mathrm{s}}\tau_{\mathrm{p}})$.

With these normalizations the rate equations (1.19) and (1.20) now assume a simple dimensionless rate equations form

$$\frac{\mathrm{d}n}{\mathrm{d}t} = J - N - NP - N, \tag{3.1}$$

$$\frac{\mathrm{d}p}{\mathrm{d}t} = \gamma NP + \beta N, \tag{3.2}$$

where $\gamma = \tau_{\mathrm{s}}/\tau_{\mathrm{p}} \sim 1,000$ for a typical laser diode, the variables J, N, P and t are now dimensionless quantities.

Assuming the applied current $J = J_0 + j(t)$ where the modulation current $j(t) = \frac{1}{2}j_1\mathrm{e}^{\mathrm{i}\omega t} + \frac{1}{2}j_1^*\mathrm{e}^{-\mathrm{i}\omega t}$. This is followed by a "perturbation analysis", so called because the analysis involves the assumption that higher harmonics are much weaker than and thus can be derived from perturbations originating from products of lower harmonic terms, as generated by the product of n and p (the stimulated emission term in the dimensionless simple rate equations (3.1) and (3.2)). This procedure gives the following results for the harmonic amplitudes [23]:

$$n_1 = j_1 g(\omega)/f(\omega), \tag{3.3a}$$

$$s_1 = \gamma j_1 (s_0 + \beta)/f(\omega), \tag{3.3b}$$

$$n_N = \frac{1}{2}\left(\sum_{i=1}^{N-1} n_i s_{N-i}\right)\left(\frac{-g(N\omega) - \gamma n_0}{f(N\omega)}\right), \tag{3.3c}$$

$$s_N = \frac{1}{2}\left(\sum_{i=1}^{N-1} n_i S_{N-i}\right)\left(\frac{-\gamma(s_0 + \beta) + \gamma h(N\omega)}{f(N\omega)}\right), \tag{3.3d}$$

where

$$g(\omega) = \mathrm{i}\omega + \gamma(1 - n_0), \tag{3.3e}$$

$$h(\omega) = \mathrm{i}\omega + (1 + s_0), \tag{3.3f}$$

$$f(\omega) = h(\omega)g(\omega) + \gamma n_0(s_0 + \beta), \tag{3.3g}$$

where γ is the ratio of the photon to spontaneous carrier lifetime ($\sim 10^3$). n_N, s_N are the coefficients of the Fourier expansion of the normalized electron and photon densities, respectively:

$$n = n_0 + \sum_k \left(\frac{1}{2}n_k \mathrm{e}^{\mathrm{i}k\omega t} + \frac{1}{2}n_k^* \mathrm{e}^{-\mathrm{i}k\omega t}\right) \tag{3.4a}$$

$$s = s_0 + \sum_k \left(\frac{1}{2}s_k \mathrm{e}^{\mathrm{i}k\omega t} + \frac{1}{2}s_k^* \mathrm{e}^{-\mathrm{i}k\omega t}\right) \tag{3.4b}$$

where $*$ represents complex conjugate. γ is the ratio of the electron to photon lifetime, and β is the familiar spontaneous emission factor. The factor $f(\omega)$ in (3.3a) and (3.3b) gives rise to the Relaxation Oscillation (RO) resonance characteristic. The Q-factor of this resonance is determined primarily by β which, apart from its formal definition as the spontaneous emission factor, can be fudged to account for other physical mechanisms such as lateral carrier diffusion [24]. The factors $f(N\omega)$ in the expressions for higher harmonics indicates that the Nth harmonic has N resonance peaks at frequencies ω_r/N, where $\omega_\mathrm{r} \sim \sqrt{\gamma(j_0 - 1)}$ is the RO frequency. The modulated output is thus especially rich in harmonic content at modulation frequencies equal to *sub-multiples* of ω_r.

Figure 3.1 shows a plot of the harmonic distortion characteristics when "prefiltering" is applied to the modulation current to compensate for the RO resonance, i.e., let $j_1 = Jf(\omega)$ in (3.3), so that the first (fundamental) harmonic response is flat. The parameters used are $\beta = 10^{-4}$, $j_0 = 1.6$, $\gamma = 2,000$, spontaneous lifetime $= 3\,\mathrm{ns}$, and the modulation depth $= 80\%$. It shows that

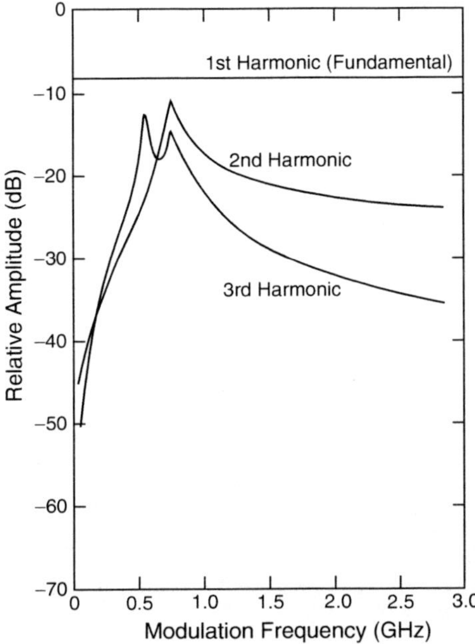

Fig. 3.1. Calculated harmonic amplitudes with prefiltering of modulation current. (From [23], ©1980 Elsevier. Reprinted with permission)

the harmonic distortion is actually worst not at the RO frequency but at submultiples of it.

Since the same factor $f(\omega)$ giving rise to the RO resonance is also responsible for the resonance peaks of higher harmonics, it follows that lasers having a high RO resonance Q-factor would have larger harmonic distortion. Indeed, this is what is observed experimentally. Figure 3.2a shows experimentally measured harmonic distortions for a proton-implant isolated stripe laser which has a relaxation oscillation resonance at about 1.7 GHz in the small signal response; the peak of which is about 8 dB above the "baseband" (low frequency) value. The data was obtained with the laser biased at $1.2\times$ threshold, driven with a sweep oscillator to an optical modulation depth of $\sim$70%. The drive amplitude is adjusted at different frequencies so that the first harmonic response is constant (i.e., prefiltering of the current). The detected output from the photodetector (rise time 100 ps) was fed into a microwave spectrum analyzer. Figure 3.2b shows a similar plot for a TJS [25] laser, which has no discernible resonance peak in the small signal response preceding the fall-off at 1.8 GHz. The distortion characteristic contrasts sharply to that of Fig. 3.2a.

While the above results show that harmonic distortions are very significant when modulated at frequencies above approximately one-third of the RO frequency, they nevertheless would not affect system performance in a significant way *if* the modulation is band-limited to below the RO frequency (i.e., low-pass filter the *received* optical signal).

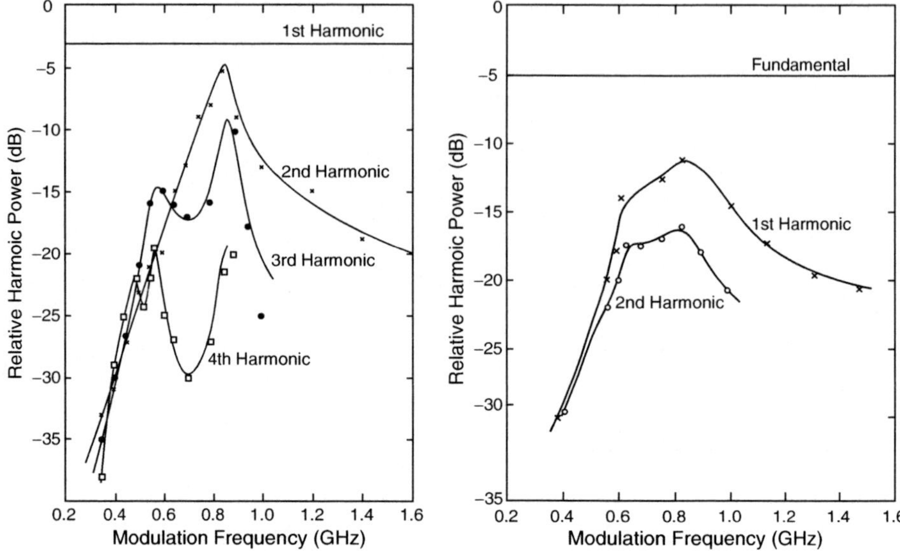

Fig. 3.2. Measured harmonic distortions of lasers (**a**) with, and (**b**) without a prominent RO resonance; prefiltering is applied to the modulation current to keep the first harmonic constant at all frequencies. (From [23], ©1980 Elsevier. Reprinted with permission)

In conclusion, an analytic approach was introduced in which the harmonic distortions in the intensity modulated output of a laser diode due to a sinusoidal modulating current was derived and compared favorably with measurements, yielding the principal result that the strength of the relaxation oscillation resonance plays a central role in harmonic distortions. These analytic results corroborate well with measured data lending credence to the analytic model, which then will form the basis of analytic studies of intermodulation distortions in Sect. 3.2.

3.2 Intermodulation Distortion

In Chap. 2 where direct modulation of semiconductor lasers was described, it was found that there are three key laser parameters of practical relevance which directly affect the modulation bandwidth, namely, the differential optical gain constant, the photon lifetime, and the internal photon density at the active region [15]. Successful tackling of each of these three parameters led to the first semiconductor lasers with a direct modulation bandwidth exceeding 10 GHz [26, 27] (Chap. 4). A direct modulation bandwidth of >10 GHz is now common place using advanced materials such as strained layer or quantum-confined media as the active lasing medium (Chap. 5). One important application for multi-GHz bandwidth semiconductor laser is multichannel (RF) frequency division multiplexed transmission of analog or microwave signals, in addition to advanced military radar and antenna systems. The most significant of these applications are successful commercial deployments of CATV distribution networks, as well as broadband cable modem Internet access via Hybrid Fiber Coaxial (HFC) cable plant. An obvious quantity of concern in these systems is the nonlinear distortion characteristics of the laser, which accounts for the lion's share of distortions in a linear (a.k.a. analog) fiber optic link. Section 3.1 introduced a perturbation analytic formulation for prediction of distortions in direct modulation of laser diodes. This section will use that formulation to predict the fundamental third order intermodulation distortion in the intensity modulated output of the laser diode. It is well known that a well behaved semiconductor laser (i.e., those with a linear light-current characteristic without kinks and instabilities above lasing threshold) exhibits very little nonlinear distortion when modulated at low frequencies (below a few tens of megahertz) [22]. This is to be expected since at such low modulation speed, the laser is virtually in a quasi-steady state as it ramps up and down along the (linear) light-current curve, and consequently the linearity of the modulation response is basically that of the cw light-current characteristic, which is excellent in well behaved laser diodes. Measurements and analysis have shown that second harmonic distortions of lower than -60 dB can be readily accomplished at the low-frequency range [27]. However, it was also shown that as the modulation frequency increases, the harmonic distortions increase very rapidly – at modulation frequencies above 1 GHz, the second and third harmonics can be as high as -15 dBc at a moderate optical modulation depth ($\sim$70%) [20,21,23]. These results can be well explained by a perturbation analytic solution of the nonlinear laser rate equations, which describes the interaction of photon and electron fluctuations – specifically the nonlinear product term of the electron and photon densities due to *stimulated emission as the origin of the large harmonic distortions observed at high frequencies*. In most *multichannel* signal transmission systems where baseband signals from different channels are modulated onto a number of well separated high-frequency carriers, second (or higher order) harmonic distortions generated by signals in a channel are actually of little concern since they generally do not fall within the frequency band where the carriers congregate unless the carriers span more than a decade in frequency. The relevant quantity of concern under those circumstances is the third order intermodulation (IM) product of the laser transmitter: two signals at frequencies ω_1 and ω_2 within a certain channel can generate intermodulation products at frequencies $2\omega_1 - \omega_2$ and $2\omega_2 - \omega_1$ may fall on another channel thus causing cross-channel interference. This is known as Intermodulation Distortion (IMD). Relevant questions include the dependence of IM products on modulation depth, signal frequencies, magnitude of relaxation oscillation, etc. These are the topics of consideration in this section.

IMD characteristics of high-speed laser diodes capable of direct modulation at multi-gigahertz frequencies have been studied both theoretically and

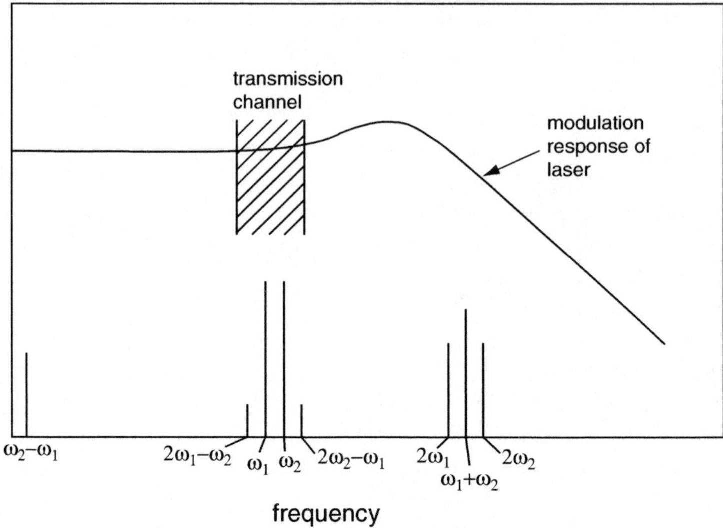

Fig. 3.3. Illustration of sidebands and harmonics generated by a two-tone modulation of a laser diode. This simulates narrowband transmission at a high carrier frequency. (From [28], ©AIP 1984. Reprinted with permission)

experimentally [28]. These results will be described in the following. The experimental study consists of modulating the lasers with two sinusoidal signals, 20 MHz apart, and observing the various sum, difference, and harmonic frequencies thus generated. The major distortion signals considered here are shown in Fig. 3.3. The principal distortion signals of practical concern, as mentioned above, are the third order IM products, at frequencies $2\omega_1 - \omega_2$ and $2\omega_2 - \omega_1$. The various distortion signals are systematically studied as one varies the frequency (ω) of the modulating signals (the two signals being ω and $\omega + 2\pi \times 20\,\mathrm{MHz}$), the optical modulation depth (OMD), and laser bias level. The OMD is defined as B/A, where B is half of the peak-to-peak amplitude of the modulated optical waveform and A is the optical output from the laser at the dc bias level. Major observed features are summarized as follow:

1. At low modulation frequencies (a few hundred MHz) all the lasers tested exhibit very low IM products of below $-60\,\mathrm{dBc}$ (relative to the signal amplitude) even at an OMD approaching 100%.
2. Second harmonics of the modulation signals increase roughly as the square of OMD, while the IM products increase as the cube of OMD.
3. The relative amplitude of the IM product (relative to the signal amplitude) increases at a rate of 40 dB per decade as ω increases, reaching a plateau at one-half of the relaxation oscillation frequency, and picks up the 40 dB per decade increment as ω exceeds the relaxation oscillation frequency. A typical value of the IM product at the plateau is $-50\,\mathrm{dBc}$ at an OMD of 50%.

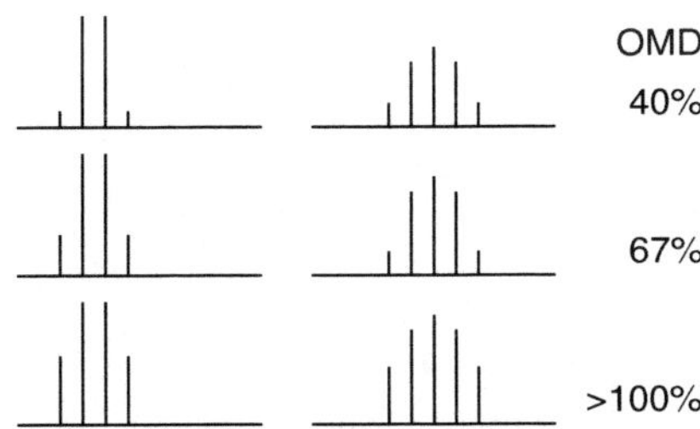

Fig. 3.4. Measured harmonics and IM products generated in a high-speed laser diode under two-tone modulation. The two tones are 20 MHz apart and centered at $\approx 3\,\mathrm{GHz}$. (From [28], ©AIP 1984. Reprinted with permission)

4. In some lasers the IM product may show a peak at one-half of the relaxation oscillation frequency. The magnitude of this peak is found to roughly correspond to the magnitude of the relaxation oscillation resonance peak in the small-signal modulation response of the laser.

Figure 3.4 shows the IM and harmonic distortions of a high-frequency laser diode under the two-tone modulation as described above, at $\omega = 2\pi \times 3\,\mathrm{GHz}$, at various OMD's. The relaxation oscillation frequency of this laser is at 5.5 GHz. The observed data of IM and second harmonic as functions of OMD and ω are plotted in the graphs in Figs. 3.5 and 3.6 respectively; the various curves in those graphs are from theoretical calculations as described below. The analytical results, based on the simplest rate equation model, can explain the above observed features very well.

The spirit of the analysis follows closely that employed in the previous harmonic distortion perturbation analysis. One starts out with the simple rate equations and assumes that the harmonics are much smaller than the fundamental signals. The photon and electron fluctuations at the fundamental modulation frequency are thus obtained with a standard small-signal analysis, neglecting terms of higher harmonics. The fundamental terms are then used as drives for the higher harmonic terms. In IM analysis where more than one fundamental drive frequencies are present, one can concentrate on the distortion terms as shown in Fig. 3.3. A "perturbation" approach can be adopted as follow: the amplitudes of the fundamental terms $(\omega_1, \omega_2) \gg$ those of second order terms $(2\omega_1, 2\omega_2, \omega_1 \pm \omega_2) \gg$ those of third order terms $(2\omega_{1,2} - \omega_{2,1})$. The perturbative analysis then follows in a straightforward manner. Denote the steady state photon and electron densities by P_0 and N_0, and the fluctuations of the photon and electron densities by lower case n and p at a frequency given by the superscript. For each of the eight frequency components shown in Fig. 3.3, the small-signal photon and electron density fluctuations are given by the following coupled linear equations:

$$\mathrm{i}\omega n^{\omega} = -(N_0 p^{\omega} + P_0 n^{\omega} + n^{\omega} + D^{\omega}), \tag{3.5a}$$

$$\mathrm{i}\omega p^{\omega} = \gamma(N_0 p^{\omega} + P_0 n^{\omega} - p^{\omega} + \beta n^{\omega} + G^{\omega}). \tag{3.5b}$$

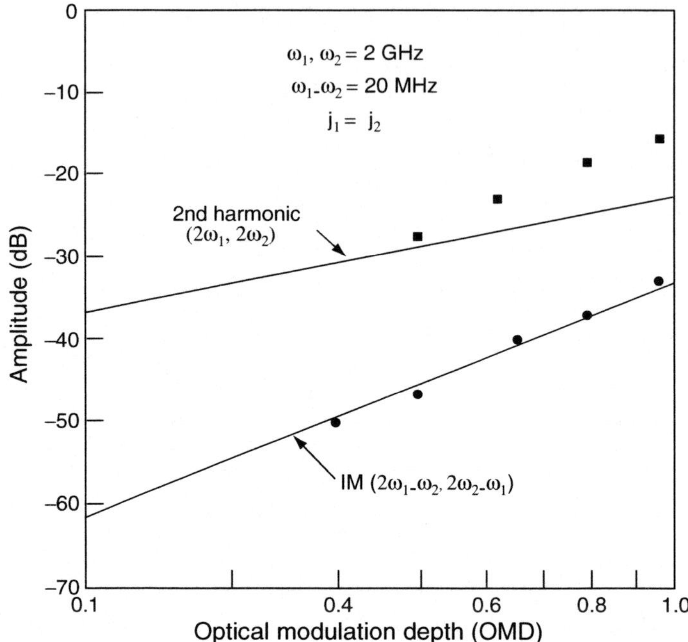

Fig. 3.5. Plots of second harmonic and third order IM amplitudes (relative to the signal amplitude) as a function of optical modulation depth (OMD), with the signal frequency at 2 GHz. Experimental data points obtained with a high-frequency laser diode are also shown. (From [28], ©AIP 1984. Reprinted with permission)

The driving terms, D^ω and G^ω, are given in Table 3.1 for each of the eight frequency components. The quantities j_1, j_2 are the modulating currents at frequencies ω_1 and ω_2. The quantities γ and β in (3.5) are the ratio of the carrier lifetime to photon lifetime and the spontaneous emission factor, respectively. The ω's in (3.5) are normalized by l/τ_s where τ_s is the carrier lifetime and the n's, p's, and j's are normalized in the usual fashion as outlined at the beginning of Sect. 3.1. One can solve for the n's and p's at each of the eight frequency components. To simplify algebra one can consider a practical case of transmission of a single channel in a narrowband centered around a high-frequency carrier, as diagrammatically depicted in Fig. 3.3. Specifically, the following is assumed:

1. $\omega_1 = \omega_c - \frac{1}{2}\Delta\omega$, $\omega_2 = \omega_c + \frac{1}{2}\Delta\omega$, $\Delta\omega \ll \omega_c$, ω_c is the center frequency of the channel.
2. $\beta \ll 1$; $1 - N_0 \sim \beta$.

The first assumption implies that the carrier frequency ω_c is much higher than $\Delta\omega$. The second assumption is based on the fact that $\beta \lesssim 10^{-3}$, and that the clamping of steady state electron density when the laser is above lasing

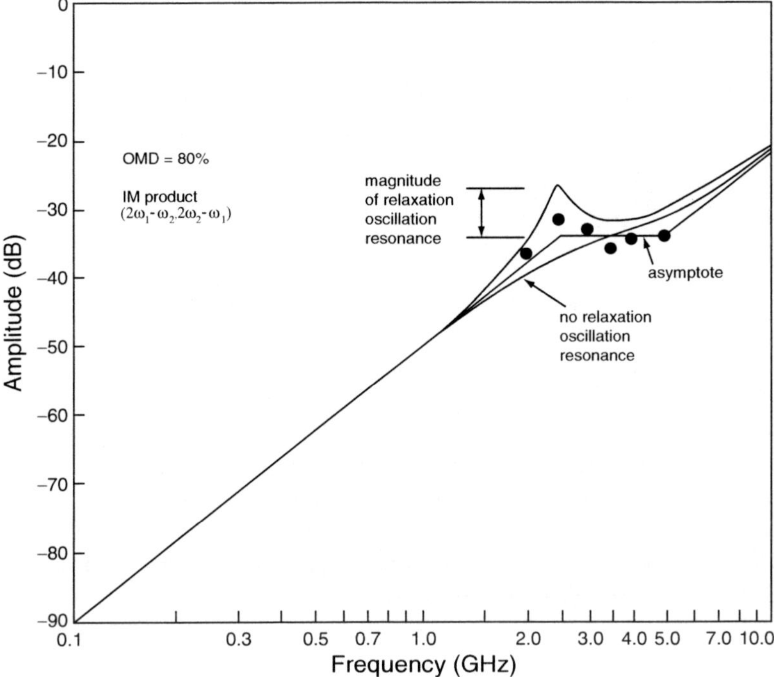

Fig. 3.6. Plots of third order IM amplitudes (relative to the signal amplitude) as a function of signal frequency, at an OMD of 80%. (From [28], ©AIP 1984. Reprinted with permission)

Table 3.1. Driving terms of the various harmonic and IM signals

ω	$D(\omega)$	$G(\omega)$
$\omega_{1,2}$	$j_{1,2}$	0
$2\omega_{1,2}$	$\frac{1}{2}n^{\omega_{1,2}}p^{\omega_{1,2}}$	
$\omega_1 - \omega_2$	$\frac{1}{2}\left[n^{\omega_1}(p^{\omega_2})^* + p^{\omega_1}(n^{\omega_2})^*\right]$	
$\omega_1 + \omega_2$	$\frac{1}{2}\left(n^{\omega_1}p^{\omega_2} + p^{\omega_1}n^{\omega_2}\right)$	Same as $D(\omega)$
$2\omega_1 - \omega_2$	$\frac{1}{2}\left[n^{2\omega_1}(p^{\omega_2})^* + p^{2\omega_1}(n^{\omega_2})^* \right.$	
	$\left. +n^{\omega_1-\omega_2}p^{\omega_1} + p^{\omega_1-\omega_2}n^{\omega_1}\right]$	
$2\omega_2 - \omega_1$	Interchange ω_1 and ω_2 above	

threshold as explained at the beginning of Chap. 2. The amplitudes of the eight frequency components of Fig. 3.3 are as follows:

$$p^{\omega_{1,2}} = j_{1,2}/f(\omega_c), \tag{3.6}$$

where $f(\omega)$ takes the form of $1 + i\omega/\omega_0 Q + (i\omega/\omega_0)^2$, where $\omega_0 = \sqrt{\gamma P_0}$ is the dimensionless (normalized to $1/\tau_s$) relaxation oscillation frequency, and Q depends on, among other things, β and the bias level;

$$p^{2\omega_{1,2}} = (i\omega_c)^2 j_{1,2}^2/\gamma P_0^2 f(2\omega_c), \tag{3.7}$$

$$p^{\Delta\omega} = ij_1 j_2^* \Delta\omega/\gamma P_0^2, \tag{3.8}$$

$$p^{\omega_1+\omega_2} = ij_1 j_2 (2\omega_c)^2/\gamma p_0^2 f(2\omega_c), \tag{3.9}$$

$$p^{2\omega_1-\omega_2} = -\frac{1}{2}\frac{ij_1^2 j_2^* \omega_c^2(\omega_c^2 + \gamma P_0)}{\gamma^2 P_0^3 f(2\omega_c)}. \tag{3.10}$$

Taking $j_1 = j_2 = j$, the relative second harmonic $(p^{2\omega_{1,2}}/p^{\omega_{1,2}})$ and intermodulation $(p^{2\omega_1-\omega_2}/p^{\omega_{1,2}})$, as given in (3.6), (3.7), and (3.10), are plotted in Fig. 3.5 as a function of the OMD $(= 2j/p_0)$, at a signal frequency of $2\,\mathrm{GHz}$ (i.e., $\omega_c/\tau_s = 2\pi \times 2\,\mathrm{GHz}$). The data points shown are obtained with a high-speed laser diode with the relaxation oscillation frequency at $5.5\,\mathrm{GHz}$. The amplitude of the IM signal (3.10) is plotted in Fig. 3.6 as a function of carrier frequency ω_c, at a fixed OMD of 80%, assuming $\omega_0/\tau_s = 2\pi \times 5.5\,\mathrm{GHz}$. The IM characteristics at other values of OMD can be obtained by shifting the curve of Fig. 3.6 vertically by an amount as given in Fig. 3.5. The values of other parameters are $\tau_s = 4\,\mathrm{ns}$, $\tau_p = 1\,\mathrm{ps}$. The actual small-signal modulation response of the lasers tested showed almost no relaxation oscillation resonance, and the value of Q was taken to be 1 accordingly. The general trend of the experimental data agrees with theoretical predictions quite well.

The above results are significant in that:

1. The linearity of the cw light-current characteristic (as well as distortion measurements at low frequencies) are *not* reliable indications of the IM performance at high frequencies.
2. Although the IM product initially increases at a rapid rate of $40\,\mathrm{dB/dec}$ as the modulation frequency is increased, it does plateau to a value of $\sim\!-45\,\mathrm{dBc}$ which is satisfactory for many applications, including, for instance, television signal transmission.

4

Direct Modulation Beyond X-Band
by Operation at High Optical Power Density

According to (2.1) the modulation bandwidth of a laser diode is proportional to the square root of the internal photon density, which is proportional to the output optical power density. For GaAs lasers commonly used in short distance data communication links, increasing the optical power density can bring about undue degradation or even catastrophic failure of the laser unless the structure of the laser is suitably designed. One common approach to raising the ceiling of reliable operating power of semiconductor lasers is by means of a large optical cavity [29]. The mechanism responsible for a higher catastrophic damage power in these devices is by lowering the optical power density at the active layer, since such damage commonly originates from the active layer near the crystal facet. This maneuver, however, serves little to increase the modulation bandwidth because the quantity of concern here, the *photon density within* the active region [P_0 in (2.1)], remains unchanged. A laser suitable for high speed operation should therefore be one with a tight optical confinement in the active region along the entire length of the laser, with a transparent window at the end regions capable of withstanding a high optical power without catastrophic damage. The use of a transparent window structure to increase the catastrophic damage level has already been demonstrated before [30, 31]. The experiments described in this chapter serve more to illustrate the basic principle of laser modulation than anything else, since in practice most lasers used for communications are of the long-wavelength type (constructed of quaternary compound semiconductors, which do *not* suffer catastrophic mirror damage at high power densities). Although they *do* suffer excessive heating effects and a concomitant drop in modulation efficiency. Hence the pathway to obtaining truly outstanding high speed behavior must be through other means. These aspects will be discussed in Chap. 5. Nevertheless the dependence of modulation speed on internal photon density can be *independently* illustrated with a "window" buried heterostructure laser fabricated on a semi-insulating substrate; this laser is basically identical to that described in Chap. 2, except for a "transparent window" region at the end facets which renders the laser immune to catastrophic mirror damage. In this

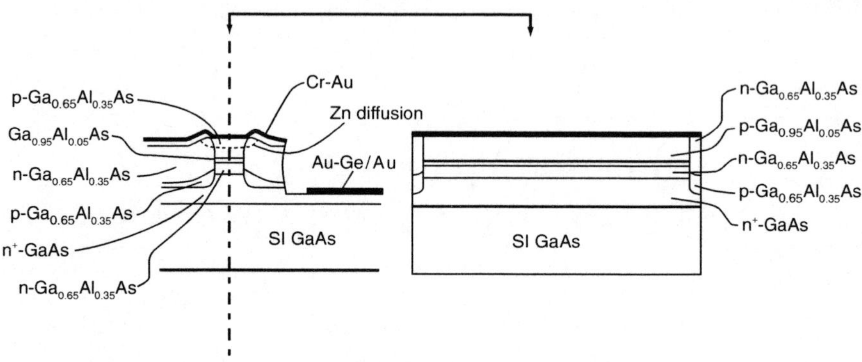

Fig. 4.1. Schematic diagram of a window buried heterostructure laser on a semi-insulating substrate. (From [27], ©1984 AIP. Reprinted with permission)

manner a direct comparison can be made, and the photon density dependence clearly illustrated. This laser served as a vehicle for the landmark demonstration [27] of the possibility of a laser diode possessing a baseband bandwidth beyond 10 GHz operating at room temperature – the figurative "four-minute mile" as far as direct laser modulation is concerned.

The laser used in this experiment is shown in schematically Fig. 4.1. The device is structurally similar to the buried heterostructure laser on semi-insulating substrate as reported in [16] (Chap. 2), except that here the end regions near the facets are covered by a layer of unpumped GaAlAs which forms a *transparent* window. A precise cleaving technique results in the facet within several microns from the edge of the double heterostructure. The optical wave propagates freely in the transparent end window region. As a result of diffraction, only a small amount of light reflected from the crystal facet couples back into the active region. This reduces the effective reflectivity of the end mirrors of the laser. The exact value of the effective reflectivity depends on the length (L) of the window region. The theoretical value of the effective reflectivity, assuming a fundamental Gaussian beam profile, is reduced to 5% for $L = 5\,\mu$m. The actual values of L for the devices fabricated lie around this value. It has been predicted theoretically [13] and demonstrated experimentally [32] that in the modulation characteristics of a laser with a reduced mirror reflectivity, the relaxation oscillation resonance is suppressed. This feature, as shown in what follows, is demonstrated by the present device.

The cw light vs. current characteristic of a window laser is shown in Fig. 4.2. The threshold current of these devices ranges from 14 to 25 mA. The threshold transition is softer than a regular laser of the same structure, which is a direct result of the reduced reflectivity as described before [13, 32]. The catastrophic damage threshold in these devices is beyond 120 mW under pulse operation. Under cw operation, the maximum operating power is limited by heating to 50 mW. The microwave modulation characteristics of the devices were measured with a standard experimental arrangement as shown

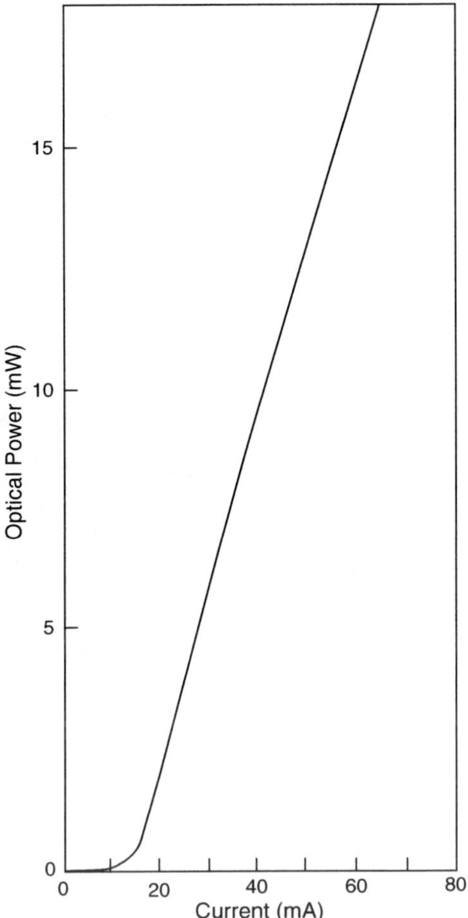

Fig. 4.2. cw light vs. current characteristics of a window buried heterostructure laser on a semi-insulating substrate. (From [27], ©1984 AIP. Reprinted with permission)

in Fig. 4.3. The photodiode used was an improved version of the on reported in [15,33] and was fully calibrated up to 15 GHz by recording the output signal on a microwave spectrum analyzer when the photodiode is illuminated by a picosecond mode-locked dye laser. The electrical system was calibrated up to 15 GHz by removing the laser and photodiode and connecting point A directly to point B as shown in Fig. 4.3. In this way, every single piece of electrical cable and connector, each of which will contribute at least a fraction of a dB to the total system loss at frequencies as high as 10 GHz, can be accounted for. The modulation data are first normalized by the electrical system calibration using a storage normalizer, and are then normalized by the photodiode response. The normalized modulation response of a window laser is shown in Fig. 4.4a, at various bias optical power levels. The conspicuous absence of the

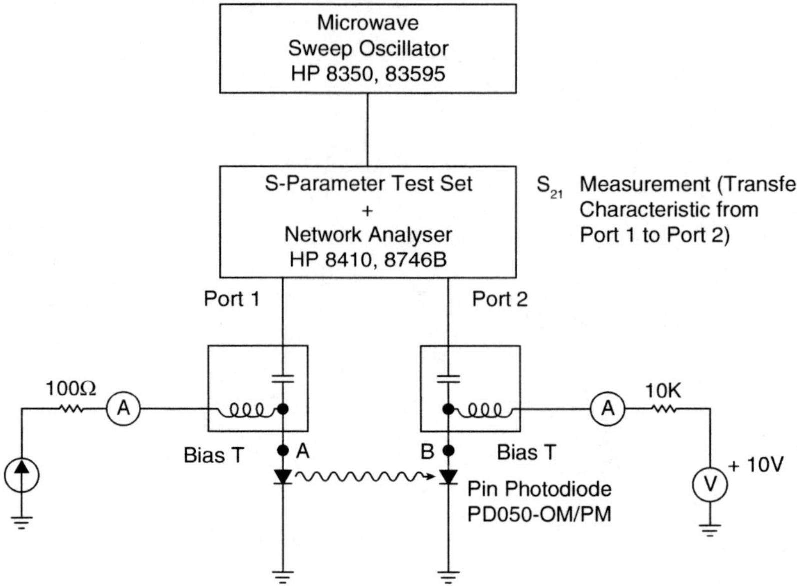

Fig. 4.3. A standard measurement system for high-frequency characterization of semiconductor lasers. (From [27], ©1984 AIP. Reprinted with permission)

relaxation oscillation peak should be contrasted with the responses of similar devices which are capable of being modulated to comparably high frequencies ($\approx 10\,\mathrm{GHz}$), examples of which are a short cavity version of the present device without a window (Chap. 2), or a regular device operating at low temperature (Chap. 5). In both of the latter instances, a strong resonance occurs when the frequency of the resonance is below ≈ 7–$8\,\mathrm{GHz}$, while the effect of parasitic elements is at least partially responsible for the reduction of the resonance amplitude at higher frequencies. The absence of relaxation oscillation in the window BH on SI lasers at all bias levels can be due to superluminescent damping effect due to the presence of the window, explained in detail in Appendix D. A plot of the $-3\,\mathrm{dB}$ modulation bandwidth of the window buried heterostructure laser against the square root of the bias optical power is shown in Fig. 4.4b. Contributions from parasitic elements are believed to be at least partly responsible for the departure of the observed data from a linear relationship at high frequencies.

In conclusion, it was demonstrated that it is fundamentally feasible to directly modulate a semiconductor laser at frequencies beyond $10\,\mathrm{GHz}$ with the laser operating at room temperature. This work, together with the experimental work described in Chaps. 2 and 5, completes the verification of the modulation bandwidth dependence on three fundamental laser parameters as given in (2.1). It is worth noting that, while the laser described in this chapter is a GaAs laser, which is:

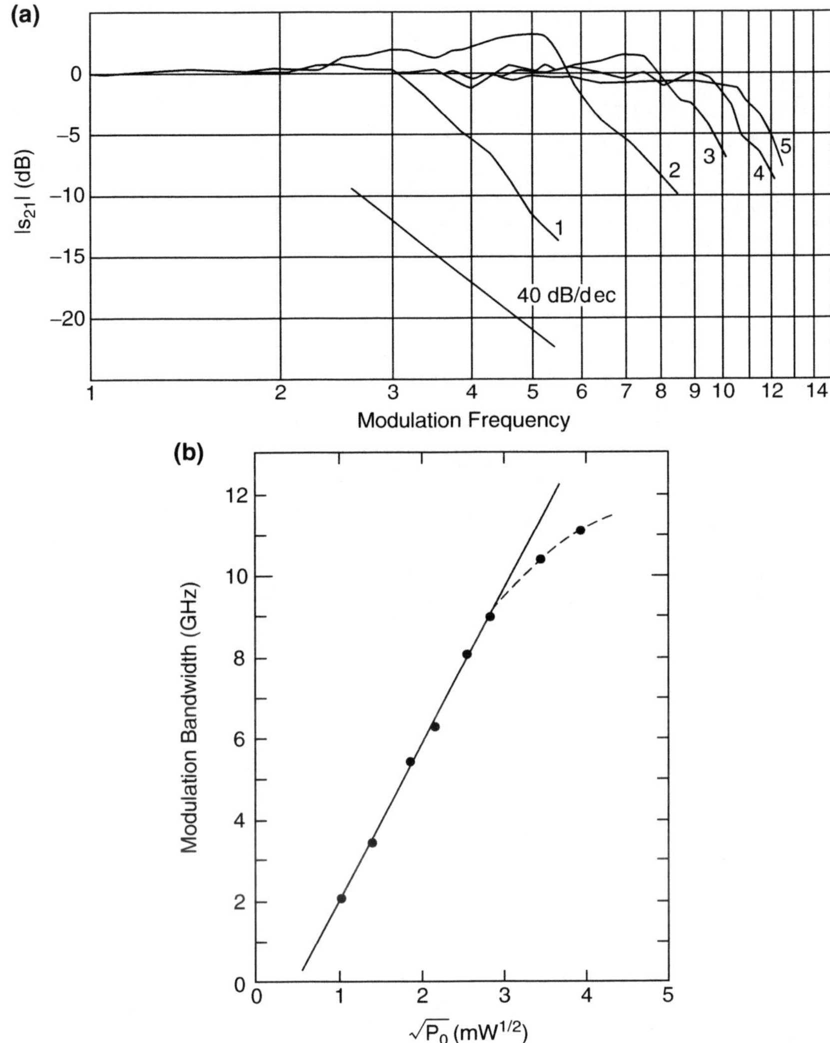

Fig. 4.4. (**a**) Modulation characteristics of a window buried heterostructure laser on a semi-insulating substrate at various bias optical power levels at room temperature. Curves 1–5 correspond to bias optical powers of 1.7, 3.6, 6.7, 8.4, and 16 mW. (**b**) The −3 dB modulation bandwidth vs. the square root of the emitted optical power. (From [27], ©1984 AIP. Reprinted with permission)

1. Not at the optimal operating wavelength for long distance fiber transmission (even though it is used for local area networks and optical interconnections between and within computers)
2. Subjected to catastrophic mirror damage common to lasers with GaAs as active region material

Standard telecommunication lasers constructed from quaternary compounds do not suffer from catastrophic mirror damage, even though the maximum operating power of those lasers are limited by thermal effects instead. All directly modulated high speed laser transmitters today, which operate in the immediate distances within the metropolitan area are constructed from quaternary semiconductor materials in the $1.3\,\mu$m wavelength region. The principles illustrated in this chapter apply well for direct modulation bandwidth limits in general.

5

Improvement in Direct Modulation Speed by Enhanced Differential Optical Gain and Quantum Confinement

Chapter 2 describes the basic intensity modulation dynamics of semiconductor lasers in general and in a most fundamental way – by proper bookkeeping of electrons and photons flowing in and out of the laser active region; the basic result is encapsulated in the very simple formula (2.1). This result, first published in 1983 [15] is of fundamental importance in studies of direct modulation properties of semiconductor lasers.

In particular, the hitherto unclear role of *differential optical gain* was clearly captured, which pointed to the fact that the direct modulation speed of laser diodes can be improved by engineering the *gain medium material properties*. However, due to the fact that the differential gain is a basic property of the gain material. The explicit dependence of the relaxation oscillation frequency on differential gain *cannot* be easily verified experimentally, since comparing different material systems may involve a multitude of factors. As an affirmative verification of the validity of this relationship, Lau et al. [26] measured the modulation bandwidth of *the same* laser diode at various low temperatures in order to increase the differential gain of the device *while keeping other material parameters and device structures unchanged* and, as such clearly demonstrated this effect. This result was further corroborated by an elegant experiment by Newkirk and Vahala [34]. These results will be illustrated in Sects. 5.1.1 and 5.1.2.

5.1 Demonstration of the Explicit Dependence of Direct Modulation Bandwidth on Differential Gain by Low Temperature Operation

5.1.1 Direct Modulation Results

Described in the following are experimental results on direct amplitude modulation of low threshold GaAs/GaAlAs buried-heterostructure lasers fabricated on semi-insulating substrates [16], operating at below room temperature.

These results show that a direct modulation bandwidth of beyond 10 GHz is attainable in laser diodes operating at modest optical power levels. However, more significant is the fact that this experiment establishes the dependence of relaxation oscillation frequency on an intrinsic laser material parameter – the differential optical gain.

The laser is mounted on a specially designed microwave package in thermal contact with a cold finger. The entire fixture resides in an enclosure in which room-temperature dry nitrogen circulates continuously (to keep out moisture). A thermocouple in close proximity to the laser records the actual operating temperature, which can be varied from $-140°$C to room temperature. The laser emission is collected by a $20\times$ objective lens from a window in the enclosure and is focused on a high-speed GaAlAs *pin* photodiode. The photodiode is an improved version of the one described in [33]; the frequency response of which was calibrated from DC to 15 GHz using a mode-locked dye laser and a microwave spectrum analyzer. The -3 dB point of the photodiode response is at 7 GHz and the -5 dB point at 12 GHz.

The light vs. current and current vs. voltage (I–V) characteristics of a 175 µm-long laser at various temperatures are shown in Fig. 5.1a,b. The lasing threshold current at room temperature is 6 mA, dropping to ≈ 2 mA at $-70°$C. The I–V curves reveal a drastic increase in the series resistance of the laser below $\approx -60°$C. This is believed to be due to carrier freeze-out at low temperatures since the dopants used, Sn (n type) and Ge (p type) in GaAlAs, have relatively large ionization energies. Modulation of the laser diode becomes very inefficient as soon as freeze out occurs because of a reduction in the amplitude of the modulation current due to a higher series resistance.

The frequency response of the lasers was measured using a sweep oscillator (HP8350) and a microwave *s*-parameter test set (HP8410, 8746). Figure 5.2 shows the response of a 175 µm-long laser at $-50°$C, at various bias levels. The responses shown here have been normalized by the *pin* photodiode frequency response. The relaxation resonance is quite prominent at low optical power levels. As the optical power is increased, the resonance gradually subsides, giving way to a flat overall response. The modulation bandwidth, taken to be the corner frequency of the response (i.e., the frequency at the relaxation oscillation peak or at the -3 dB point in cases when it is absent), is plotted against the square root of the emitted optical power ($\sqrt{P}$) in Fig. 5.3, at room temperature and at $-50°$C and $-70°$C.

Since, according to (2.1), the corner frequency is directly proportional to $\sqrt{A}$ where A is the differential optical gain, the relative slopes of the plots in Fig. 5.3 thus yield values for the relative change in A as the temperature is varied. The ratio of the slope at $22°$C to that at $-50°$C is 1.34 according to Fig. 5.3. This factor is fairly consistent (between 1.3 and 1.4) among all the lasers tested, even including those from different wafers. According to these measurements it can be deduced that the intrinsic differential optical gain of GaAs increases by a factor of ≈ 1.8 by cooling from $22°$C to $-50°$C assuming that the photon lifetime does not change with temperature. To check whether

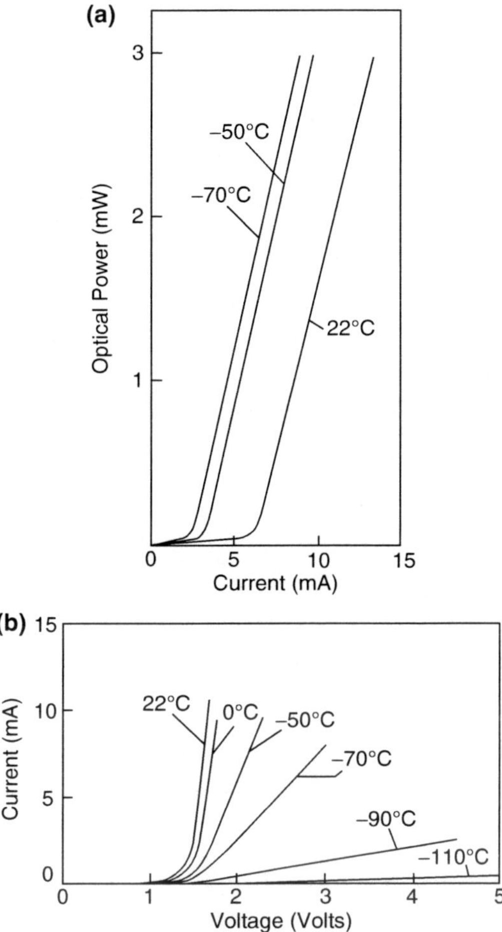

Fig. 5.1. (a) Light vs. current characteristics of a 175 µm laser at various temperatures; (b) I–V characteristics of the same laser. (From [26], ©1984 AIP. Reprinted with permission)

this result is consistent with previously calculated values, Fig. 3.8.2 in [35] can be consulted, in which the calculated optical gain is plotted against the carrier density for various temperatures. The differential gain coefficient A is the slope of the gain vs. carrier concentration plots. From these theoretical results, the ratio of A at 160 K to that at 300 K is 2.51. A simple linear interpolation yields an increase by a factor of 1.87 for A at 223 K (−50°C) over that at 300 K. This is consistent with the value obtained from the modulation measurements described above.

While the above experiment clearly demonstrates the dependence of relaxation oscillation frequency on differential optical gain, increased junction

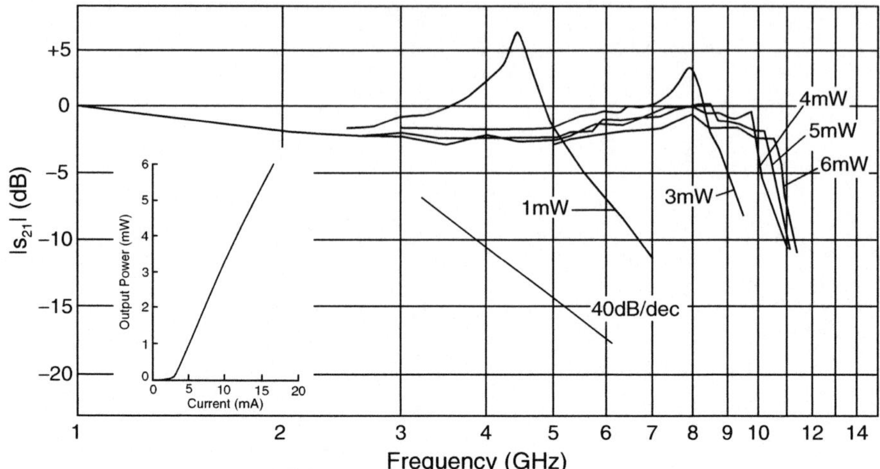

Fig. 5.2. Modulation response of a $175\,\mu$m buried-heterostructure laser on semi-insulating substrate operating at $-50°$C. (From [26], ©1984 AIP. Reprinted with permission)

resistance at low temperatures introduces very large electrical parasitic effects which could camouflage improvements in modulation response. In the next section, an elegant parasitic-free modulation method will be described which circumvents this limitation and *unequivocally* proves the explicit dependence of relaxation oscillation frequency on differential optical gain. This constitutes the basis of understanding of superior high speed direct modulation properties of advanced lasers such as quantum confined and strained layer lasers.

A very important aspect of the lasers used in the above experiments is their fabrication on semi-insulating substrates, which substantially lowers the parasitic capacitance of the laser – which has been shown to be the most damaging parasitic element in high-frequency modulation [36]. In the lower GHz range it is a general and consistent observation that the modulation response of these lasers does not exhibit any dip as observed in other lasers [12]. Measurements of the electrical reflection coefficient (s_{11}) from the laser gave indications that effects due to parasitic elements are appreciable at modulation frequencies above 7 GHz. This can account for the absence of a resonance peak in the modulation response at high optical powers (Fig. 5.2) and the slight discrepancy between the measured and the predicted at the high-frequency end (Fig. 5.3). The importance of minimizing parasitic elements by suitable laser design in attempting modulation at frequencies as high as 10 GHz cannot be overstated.

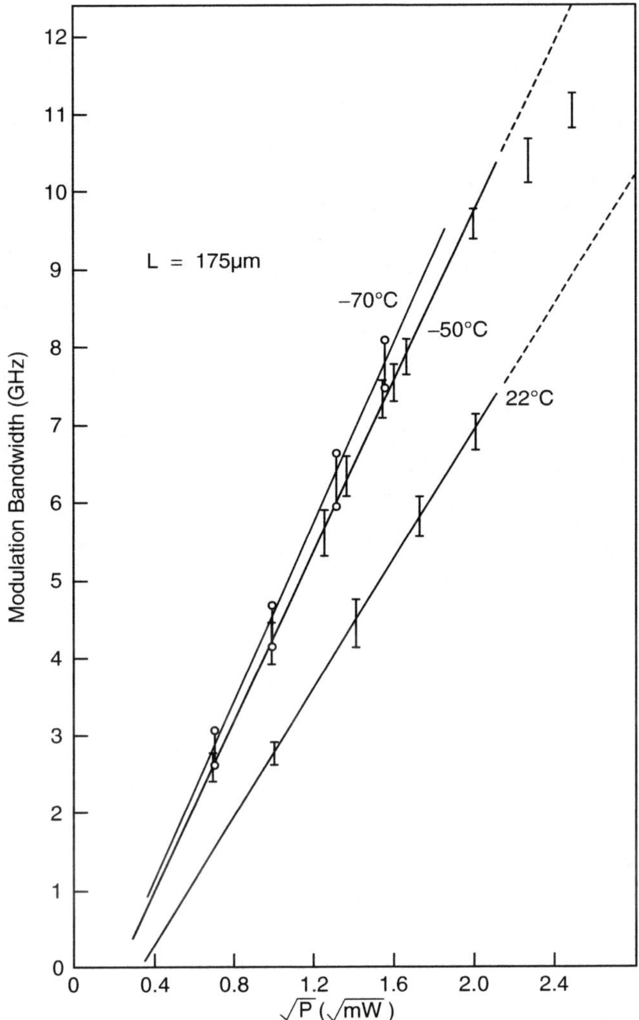

Fig. 5.3. Variation of modulation bandwidth (corner frequency of the modulation response) with the square root of the emitted optical power $\sqrt{P}$. (From [26], ©1984 AIP. Reprinted with permission)

5.1.2 Parasitic-Free Photo-Mixing Modulation Experiment

While the low temperature experiments described in Sect. 5.1.1 clearly demonstrates the dependence of relaxation oscillation frequency on differential optical gain, the high series resistance encountered at low temperature impeded collection of clean data. This problem was circumvented by a subsequent, elegant modulation technique demonstrated by Newkirk and Vahala [34] which involves *directly* modulating the carrier density in the active region of the laser

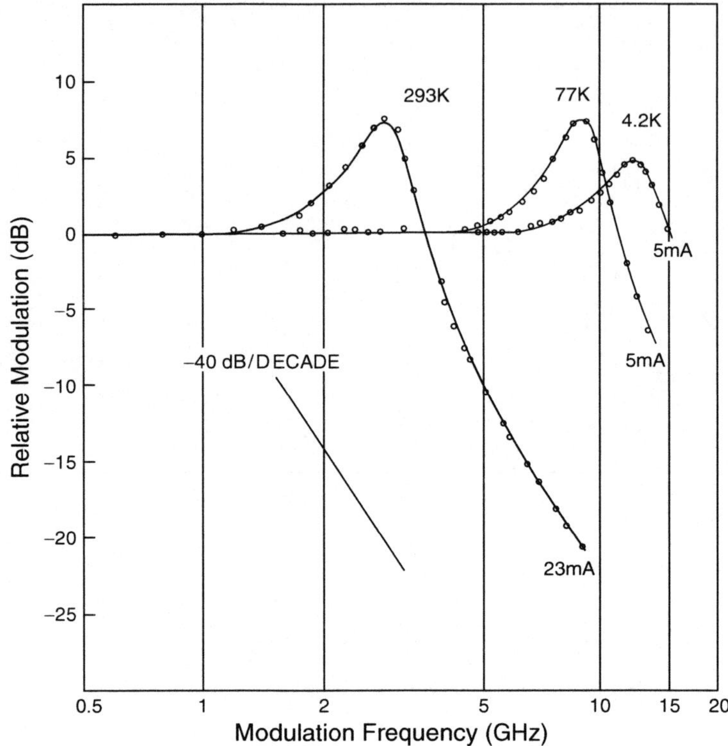

Fig. 5.4. Measured modulation response by parasitic-free photo-mixing technique at three temperatures. (From [34], ©1989 AIP. Reprinted with permission)

diode by illuminating the active region of the test laser diode with two cw laser beams which are slightly detuned (and continuously tunable) in their optical frequencies. The carriers in the active region of the test laser diode are thus directly modulated at the difference frequency of the two illuminating beams. which can be varied over an extremely wide range unimpeded by parasitic effects induced by a high series resistance at low temperature. The modulation response data shown in Fig. 5.4 [34] is thus extremely clean and approaches ideal; these data obtained at temperatures down to liquid helium temperature validates convincingly the theoretical result for direct modulation bandwidth of laser diodes (2.1).

While operating a laser diode at close to liquid nitrogen [26] or liquid helium [34] temperatures as described above clearly illustrates the basic physics of high speed modulation behavior of semiconductor lasers, it is obviously not practical to do so under most circumstances. To this end a laser diode capable of high speed operation under "normal" conditions is much desirable (Fig. 5.5). This calls for advanced materials which possess a high differential optical gain at room temperature. Two such existing examples are quantum confined

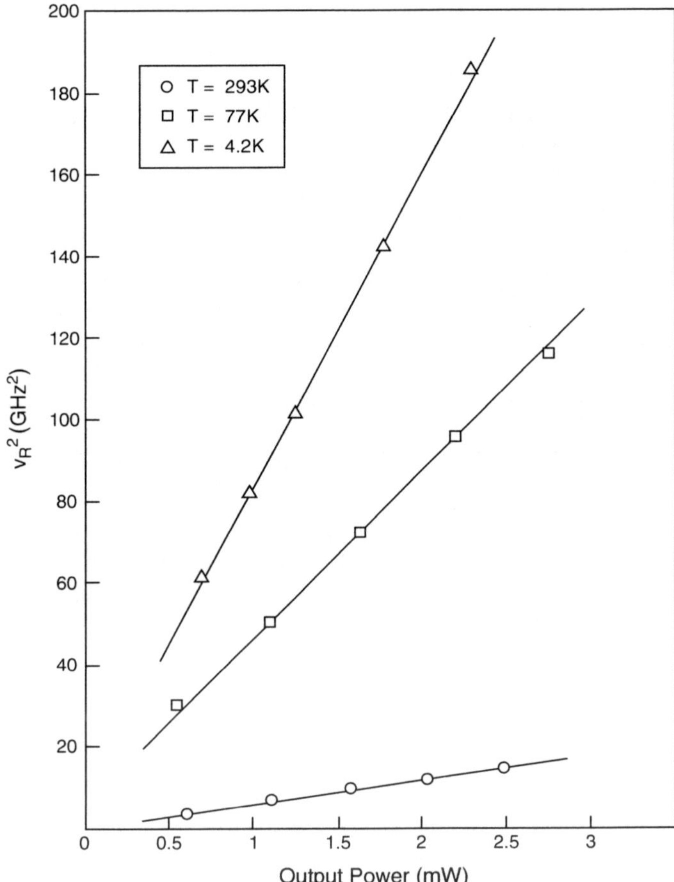

Fig. 5.5. Square of resonance frequency vs. output power. (From [34], ©1989 AIP. Reprinted with permission)

media and strained layer medium [37]. Section 5.2 briefly describes investigations in quantum confined media by Arakawa, Vahala, and Yariv who first predicted theoretically and articulated clearly [38] that quantum confinement could play an enabling role in enhancing differential gain. Additionally, the impact of quantum-confinement on the α-factor, which is inversely proportional to differential gain, was clearly identified by these authors and was shown to dramatically improve other dynamic properties such as FM/IM ratio (chirp) which determines spectral purity under modulation for single frequency lasers (such as a distributed feedback lasers prevalent in telecommunication these days). This again affirmed the critical importance of achieving high differential gain for high-performance lasers. Interested readers are referred to [37] for a parallel treatment regarding strained layer laser physics and performances.

5.2 Attainment of High Modulation Bandwidths Through Quantum-Confined Materials

The relation (2.1) clearly illustrates the significant role optical gain (or differential optical gain – to be precise, plays in the modulation speed of a laser diode. The differential optical gain in a quantum confined medium (a.k.a. lower dimensional material) can be increased significantly over that of a bulk semiconductor material. The optical gain in a medium is directly related to electron (hole) occupation of available states in the material, the latter, known as "density of states" ("DOS") is significantly different in three-dimensional (3D) from that of 2D, 1D or 0D. 3D materials do not confine the motion of electrons or holes in any direction and are popularly known as "bulk materials"; 2D materials confine the motion of electrons or holes in a plane and are popularly known as "quantum well (QW) materials"; 1D materials confine the motion of electrons in a line and are popularly known as "quantum wire" (Q-Wi) materials; 0D materials do not allow kinetic motion of the electrons at all and are popularly known as "quantum dot" (QD) materials.

It should be noted that, while quantum well lasers [39] were an area of intense interest and investigation for quite some time. The role of its DOS in enhancing dynamical properties had not been addressed coherently before the publication of [38] which predicted the effect of quantum confinement on the enhancement of relaxation oscillation frequency f_r and the reduction of α-factor, the latter a measure of wavelength "chirping" under direct modulation. The principal result of this work is summarized in Fig. 5.6 in which f_r and α are plotted as a function of wire width for a quantum wire medium, as

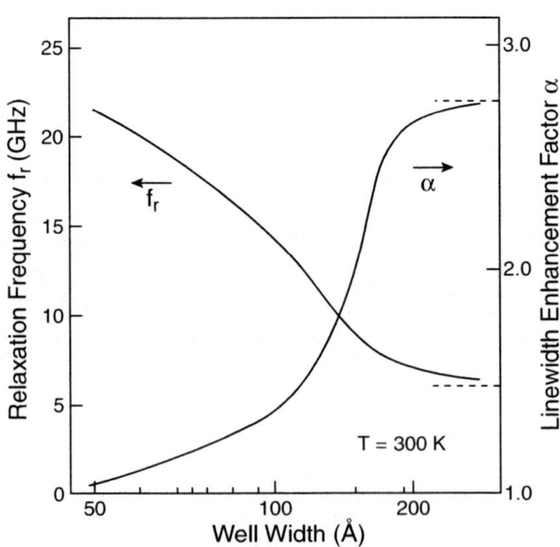

Fig. 5.6. Predicted α and f_r as a function of wire width for a quantum wire laser. (From [38], ©1984 AIP. Reprinted with permission)

the wire width is reduced the modification in density of states produces an enhancement in f_r and a reduction in α. This important publication pointed the way to continued improvements beyond quantum wells with quantum wires and dots, an area that is now poised for high impact for the same reasons.

While the superior dynamic characteristics of lower-dimension (1-D or 0-D) quantum confined lasers have been predicted as described in [38] (Fig. 5.6) their realization require fabrication of structures of sizes comparable to the electronic wave-function in order to provide quantum confinement of electrons and holes in 1, or 3 dimensions. These structures are difficult to fabricate due to their small physical dimensions; it was only relatively recently that Q-Wi (1-D) and QD (0-D) lasers have been realized, and then still not yet operating at the performance level where consistent high speed modulation data could be taken. So far measurements on these low-dimensional quantum confined lasers have been confined to lasing thresholds, spectra and their temperature dependences only. An early *independent* verification of the beneficial dynamic effects of low-dimensional quantum confinement was carried out by Arakawa, Vahala, Yariv and Lau [40] in which quantum confinement was created experiment ally with a large magnetic field for *quantum-wire behavior due to quantized cyclotron orbits*. This experiment took on a similar flavor as the earlier low temperature experiments by Lau and Vahala (Sects. 5.1.1 and 5.1.2), in that the *same* device was used to demonstrate the intended effects *without* having to account for uncertainties from comparing experimental results of devices prepared from different material systems prepared in different apparatus. This experiment indeed was shown to produce the density of states modifications associated with quantum confinement, in turn yielding the associated enhancements in differential gain, that in turn produced the predicted, and experimentally demonstrated, enhancements in modulation bandwidth, as shown in Fig. 5.7 [40]. It is thus expected that continued improvements in nano-fabrication technologies will result in consistent and reliable low-dimensional quantum-confined lasers which can bring about practical high frequency directly-modulated broad-band optical transmitter, capable of operating in the millimeter wave frequency range without resorting to narrow-band schemes such as resonant modulation, to be described in Part II of this book.

While concerted research efforts are underway to fabricate reliable high performance Quantum Wire and Quantum Dot lasers, Quantum Well lasers have already been realized for some time and now dominate the telecommunications laser market in long and intermediate reach systems. This is due to the most important and fundamental benefits relating to the dynamical properties enabled by the high differential gain (dg/dn) of these lasers, resulting in a high modulation bandwidth at modest powers and low chirp under modulation. This lends credence to the prediction (and hence the excitement) that Quantum Wire and Dot lasers will assume similar dominant positions when their fabrication and production become mature.

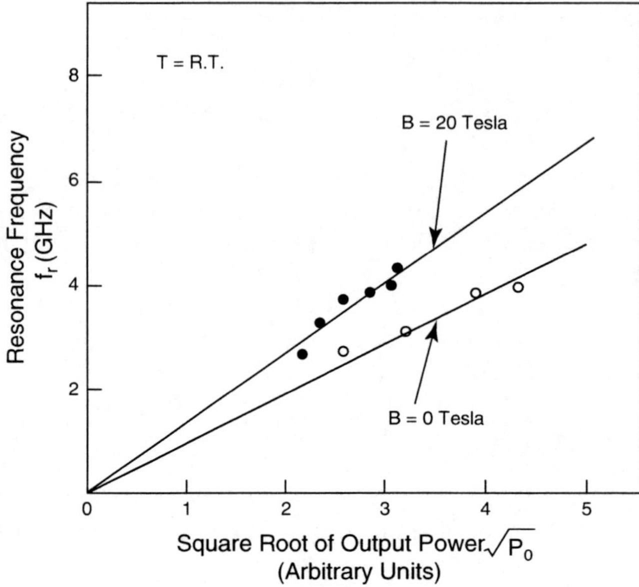

Fig. 5.7. Variation of relaxation oscillation frequency (modulation bandwidth) of a standard DH laser in a magnetic field which provides 1D quantum confinement (Q-Wi) effect. (From [40], ©1985 AIP. Reprinted with permission)

6

Dynamic Longitudinal Mode Spectral Behavior of Laser Diodes Under Direct High Frequency Modulation

6.1 Introduction

The steady state longitudinal mode spectrum of semiconductor lasers has been extensively studied, and major observed features can be understood in terms of modal competition in a common gain reservoir. It was generally agreed that gain saturation in semiconductor lasers is basically homogeneous. Thus, a well-behaved index-guided laser should oscillate predominantly in a single-longitudinal mode above lasing threshold [41, 42]. This has been verified extensively in semiconductor lasers of many different structures. It was also recognized that a single-mode laser will not remain single-mode during turn-on transients and high-frequency modulation. This can be predicted theoretically from numerical solutions of the multimode rate equations [43]. The optical spectrum of a semiconductor laser during excitation transient has been observed by many researchers [44–50]. It was generally observed that when a laser is biased at a certain dc current and excited by a current pulse, the relative amplitude of the longitudinal modes at the beginning of the optical pulse is essentially identical to the prepulse distribution. Depending on the laser structure, it will take $\sim$0.5–5 ns for the laser to redistribute the power in the various longitudinal modes to that corresponding to the CW spectrum at the peak of the current pulse. A simple analysis [51] showed that during switch-on of lasing emission, the ratio of the power in the ith mode to that in the jth longitudinal mode is given by

$$\frac{s_i(t)}{s_j(t)} = \frac{s_i(t=0)}{s_j(t=0)} \exp(G_i - G_j)t, \tag{6.1}$$

where $G_i = g_i \alpha$ is the optical gain of the ith mode, g_i is commonly represented by a Lorentzian distribution

$$g_i = \frac{1}{1 + bi^2} \tag{6.2}$$

and α is the gain of mode 0, which is assumed to be at the peak of the optical gain curve. In common semiconductor laser devices where the entire gain spectrum spans over several hundred angstroms, the value of b, which is a measure of the amount of mode selectivity, is quite small and is on the order of 10^{-4}, according to (6.1), leads to a long time constant for the different modes to settle to their steady amplitudes. An approximation used in deriving (6.1) is that spontaneous emission is neglected. It should be noted that within this approximation, (6.1) is applicable regardless of whether the total photon density undergoes relaxation oscillation or not, as can be seen from a careful examination of its derivation [51].

While (6.1) gives a fairly good description of the spectral behavior of lasers under step excitation, it cannot be used when the modulation current takes on a form other than a step. The reason is that in deriving (6.1), spontaneous emission was omitted and consequently as $t \rightarrow \infty$, it predicts that only one mode can oscillate and the amplitudes of all other modes decay to zero, regardless of starting and final pumping conditions. Therefore, it cannot be used to describe the lasing spectrum of a laser modulated by a series of current pulses. Moreover, it cannot be used to explain the lasing spectrum of a laser under high-frequency continuous microwave modulation. Previous experiments have shown that when microwave modulation is applied to an otherwise single-mode laser, the lasing spectrum will remain single-mode unless the optical modulation depth exceeds a certain critical level [52]. There was no systematic experimental study of how that critical level depends on the properties of the laser diode and modulation frequency; neither was there an analytical treatment of the phenomenon.

The purpose of this chapter is twofold: first, to present experimental results of a systematic study of the conditions for an otherwise single-mode laser to turn multimode under high frequency microwave modulation, and secondly, to develop a theoretical treatment which can explain these results, and provide a general understanding of the time evolution of lasing spectrum through simple analytical results. In addition to an increase in the number of lasing modes, it has also been observed that the linewidth of the individual lasing modes increases under high-frequency modulation [53]. This has been explained by time variation of electron density in the active region, with a concomitant variation of the refractive index of the lasing medium thus causing a shift in the lasing wavelength. This will be further explained in detail in Sect. 6.7.

6.2 Experimental Observations

The longitudinal spectrum of a laser under direct modulation obviously depends on the amount of mode selectivity in the laser. Those lasers with a built-in frequency selective element, such as that in a distributed feedback-type laser, can sustain single-mode oscillation even under turn-on transients and high-frequency modulation [54,55]. The same is true for a laser with a very short cavity length, where the increased separation between the longitudinal modes results in a larger difference in the gain of adjacent modes [48, 49], or in a composite cavity laser where additional frequency selectivity arises from intracavity interference [56]. Experimental work on lasing spectral transients under step or pulse excitation has been fairly well documented. In experiments described in this chapter, the main concern is the time-averaged lasing spectrum of lasers of various cavity lengths under high-frequency *continuous* microwave modulation at various frequencies and modulation depths. The lasers used are index-guided lasers of the buried heterostructure type with a stable single transverse mode. The CW characteristics of a 120 μm long laser are shown in Fig. 6.1. These lasers have an extremely low threshold, less than 10 mA, and display a single-longitudinal mode at output powers above ∼1.3 mW. The CW characteristics of a regular laser whose cavity length is 250 μm are shown in Fig. 6.2. The light vs. current characteristic is essentially similar, except for the higher threshold current, to that of the 120 μm laser. The longitudinal mode spectrum of this laser becomes single-mode at an output power slightly above 1 mW. The fraction of power contained in the dominant lasing mode is higher in the short laser than the long one at all corresponding output power levels. However, it should be mentioned that

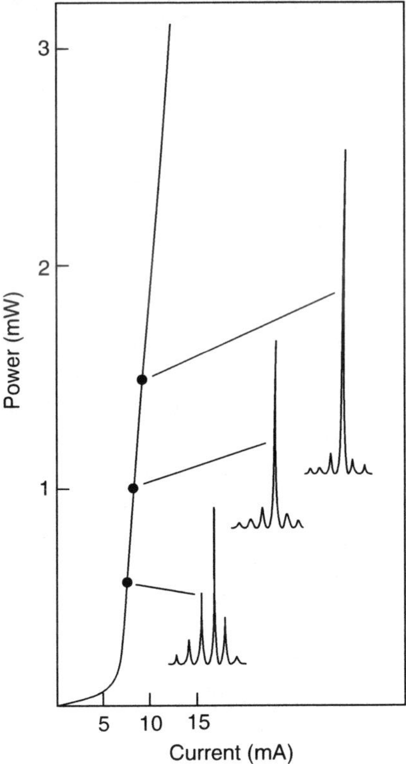

Fig. 6.1. CW light vs. current and spectral characteristics of a GaAs laser whose cavity length is 120 μm

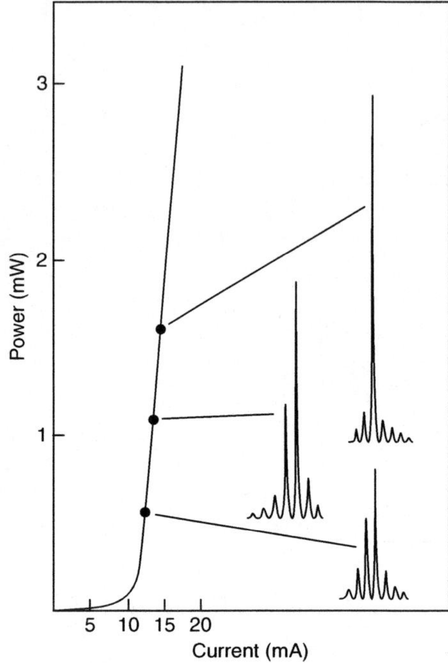

Fig. 6.2. CW light vs. current and spectral characteristics of a GaAs laser identical to that shown in Fig. 6.1 except that cavity length is $250\,\mu m$

this is only a general observation, and exceptions where a long laser has a purer longitudinal mode spectrum compared to a short laser do exist. Thus, in high-frequency modulation experiments described below, the comparison is not as much between short and long lasers as between lasers with intrinsically different mode selectivities. In any case, it has been generally observed that during turn-on transients, a short cavity laser settles to single-mode oscillation considerably faster than a long cavity laser [48,49]. Under high-frequency continuous microwave modulation with the laser biased above threshold, the increase in the number of longitudinal modes is expected to be smaller in short lasers. This is a general observation in these experiments. All the lasers tested retain their single-mode spectrum until the optical modulation depth exceeds $\sim$75–90%, depending on the purity of the original CW lasing spectrum. The optical modulation depth η is defined as the ratio of the amplitude to the peak of the modulated optical waveform [i.e., if the optical waveform is $S_0+S_1\cos\omega t$, then $\eta = 2S_1/(S_0+S_1)$]. Another interesting observation is that, contrary to common belief, this critical modulation depth does not depend on modulation frequency, at least within the frequency range of 0.5–4 GHz. Results obtained with the short laser in Fig. 6.1 are shown in Fig. 6.3a, which depicts the time-averaged spectrum at various modulation depths and frequencies between 1 and 3 GHz. A single-mode spectrum can be maintained at

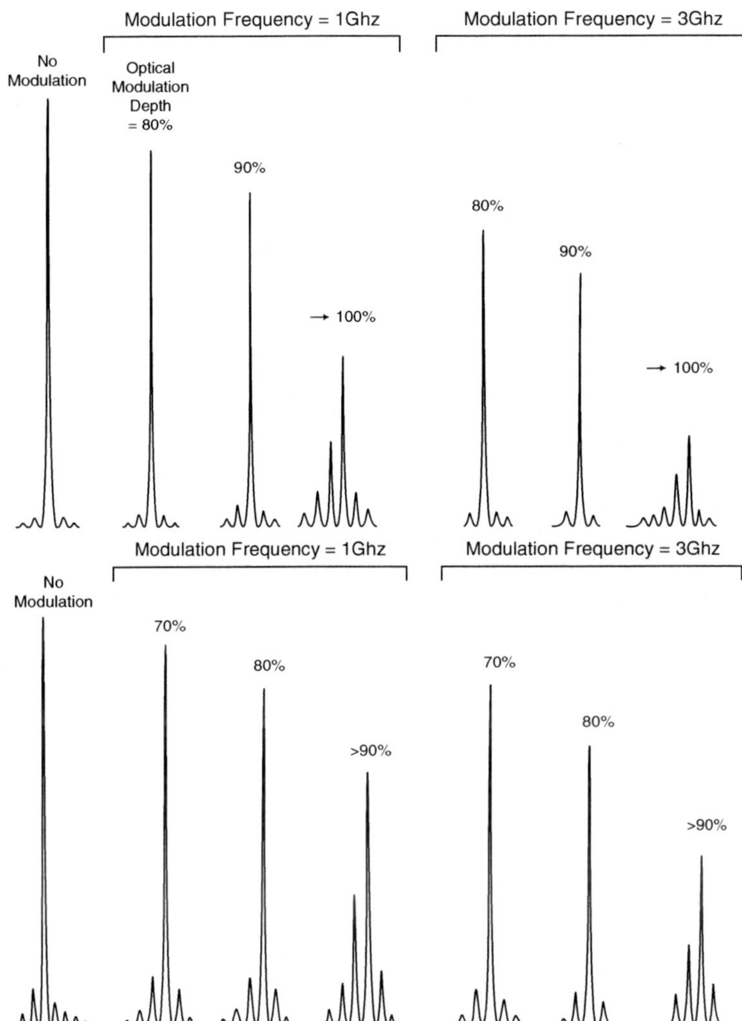

Fig. 6.3. (a) Observed time-averaged spectrum of the laser shown in Fig. 6.1 under microwave modulation at various optical modulation depths, at modulation frequencies of 1 and 3 GHz. The laser is biased at a dc optical power of 1.5 mW. (b) Same experiment as in (a) but for the laser shown in Fig. 6.2

a modulation depth up to 90% regardless of modulation frequency. The laser is biased at a dc output power of 1.5 mW. However, it can be observed that the width of the individual modes broadens at higher frequencies, although the relative amplitudes of the modes do not change. This, as mentioned before, arises from fluctuations in the refractive index of the cavity as a result of fluctuation in carrier density. A simple singlemode rate equation analysis shows that under a constant optical modulation depth, the fluctuation in carrier density increases with increasing modulation frequency, and consequently the line broadening effect is more visible at high frequencies [53]. Figure 6.3b shows a set of data similar to that in Fig. 6.3a, for the longer cavity laser whose CW characteristics are shown in Fig. 6.2. The laser is biased at an identical output power of 1.5 mW as above. Multimode oscillation occurs at a lower optical modulation depth of 75%. This is also relatively frequency independent.

Extreme care must be taken in determining the exact value of the optical modulation depth, especially at high frequencies. The drop-off in the photodetector response at high frequencies can be taken into account by precalibrating the photodiode response using picosecond pulse techniques. However, most photodiodes display an excess dc gain of several decibels, and it is very difficult to calibrate this excess gain by common picosecond pulse techniques, yet it is very important that this excess gain be taken into account when trying to determine the optical modulation depth from the observed dc and RF photocurrents. One way to do this is to observe the photodetector output directly in the time domain (oscilloscope) while modulating the laser at a "low" frequency (say, a few hundred megahertz where the photodiode response is flat) and increasing the modulation current to the laser until clipping occurs at the bottom of the output photocurrent waveform from the photodiode. This indicates clearly the level corresponding to zero optical power. The excess dc gain over the midband gain of the photodiode can then be accurately determined from the observed dc and RF photocurrents at the point of clipping.

6.3 Time Evolution Equations for Fractional Modal Intensities

As mentioned in Sect. 6.1, meaningful theoretical analysis on spectral dynamics must include both cases of positive and negative step transitions. These give insights in cases of practical interest such as that when the laser is modulated by a pseudorandom sequence of current pulses or by a continuous microwave signal. Analytic solutions are difficult to come by due to the complexity of the coupled nonlinear multimode equations which do not lend themselves to easy analytical solutions. Numerical analysis of the multimode rate equations has been previously reported for some specific cases [44, 57]. The intention of this and the following section are to derive a simple analytical solution which will allow significant insights into the problem of the time evolution of the spectrum and its dependence on various device parameters and pump conditions.

In previous analyses of laser dynamics, one sets out to find the optical response given a certain modulation current waveform. There is no easy solution to the problem through this approach even when only one longitudinal mode is taken into account, except in the limit of small signal analysis where the equations are linearized. In the case where many modes are taken into account and nonlinear effects are what one is looking for, the analysis becomes hope-lessly complex. A different approach is used here, where one asks the following question: given that the total optical output from the laser takes on a certain modulated waveform, how does the spectral content of this output vary as a function of time?

The rate equation governing the time evolution of the number of photons in the ith longitudinal mode reads

$$\frac{\mathrm{d}s_i}{\mathrm{d}t} = \frac{1}{\tau_\mathrm{p}}\left[(\Gamma g_i n - 1)s_i + \Gamma \beta_i n\right],\tag{6.3}$$

where n is the electron density normalized by $1/A\tau_\mathrm{p}$, s_i is the photon density in the ith mode normalized by $l/A\tau_\mathrm{s}$, A is the (differential) optical gain constant, β_i is the spontaneous emission factor for the ith mode, τ_p and τ_s are the photon and spontaneous lifetimes, Γ is the optical confinement factor, and g_i is the Lorentzian gain factor in (6.2) where mode 0 is taken to be at the center of the optical gain spectrum. It follows, from the constant proportionality between the stimulated and spontaneous emission rate into a mode, that $\beta_i = \beta \times g_i$ where $\beta = \beta_{i=0}$. The normalized electron density n is clamped to a value very close to l/Γ under steady state operation, and numerical computations have shown that it does not deviate significantly ($<$parts in 10^2) from that value even during heavy optical transients at high frequencies [51]. The reason that n cannot be simply taken as a constant in solving (6.3) is that the quantity $1 - n\Gamma g_i$, though small, cannot be neglected in (6.3). Let $S = \sum s_i$ be the total photon density summed over all modes. The rate equation for S is

$$\dot{S} = \frac{1}{\tau_\mathrm{p}}\left(\sum_i s_i \Gamma g_i n - S + \Gamma n \sum_i \beta_i\right).\tag{6.4}$$

Now, let $\alpha_i = s_i/S$ be the fraction of the optical power in the ith mode. The rate equation for α_i can be found from (6.3) and (6.4)

$$\dot{\alpha}_i = \frac{S\dot{s}_i - s_i\dot{S}}{S^2} = \frac{1}{\tau_\mathrm{p}}\Gamma\left(\alpha_i \sum_j (g_i - g_j)\alpha_j - \frac{\alpha_i}{S}\sum_j \beta_j + \frac{\beta_i}{S}\right)n,$$

$$i = -\infty \to \infty.\tag{6.5}$$

The normalized electron density n can now be taken as l/Γ since it appears only by itself in (6.5). The quantity $\sum_j (g_i - g_j)\alpha_j$ in (6.5) obviously depends on the instantaneous distribution of power in the modes and causes considerable difficulty in solving (6.5) unless some approximations are made. However, one can first look at the case of a laser which has only two modes (or three modes symmetrically placed about the peak of the gain curve). The exact analytical solution can then be obtained which yields considerable insight into transient modal dynamics. It should be noted at this point that, with a simple and reasonable assumption, the solution of (6.5) in the many-mode case is very similar to that in the two-mode case. Therefore, implications and conclusions drawn from the two-mode solution are directly transferable to the full many-mode solution.

6.4 A Two-Mode Laser

If only two modes exist, then it is obvious that the fraction of power contained in the two modes, α_1 and α_2, is related by

$$\alpha_1 + \alpha_2 = 1. \tag{6.6}$$

Thus, from (6.5), the time evolution equation for α_1 is

$$\dot{\alpha}_1 = \frac{1}{\tau_{\mathrm{p}}} \left[\alpha_1 \alpha_2 (g_1 - g_2) + \frac{\beta}{S}(1 - 2\alpha_1) \right], \tag{6.7}$$

where $\beta_1 = \beta_2$ is taken since they differ only by parts in 10^4 [44]. Combining (6.6) and (6.7), one has

$$\dot{\alpha}_1 = \frac{1}{\tau_{\mathrm{p}}} \left[\alpha_1 (1 - \alpha_1)\delta g + \frac{\beta}{S}(1 - 2\alpha_1) \right],$$
$$\delta g = g_1 - g_2 \,. \tag{6.8}$$

Now, one can assume that the modulated waveform of the total photon density S is that of a square wave as shown in the top parts of Fig. 6.4, so that in solving (6.8) S takes on alternate high and low values as time proceeds. A straightforward integration of (6.8) yields a solution within each modulation half cycle

$$\alpha_1(t) = \frac{1}{\tau \mathfrak{B}} \left\{ \frac{\mathfrak{A} + 2\mathfrak{B}\alpha_1(0) + \frac{2}{\tau}\tanh\left(\frac{t}{\tau}\right)}{[\mathfrak{A} + 2\mathfrak{B}\alpha_1(0)]\tanh\left(\frac{t}{\tau}\right) + \frac{2}{\tau}} \right\} - \frac{\mathfrak{A}}{2\mathfrak{B}}, \tag{6.9}$$

where

$$\frac{1}{\tau} = \frac{1}{2\tau_{\mathrm{p}}} \sqrt{\delta g^2 + 4\left(\frac{\beta}{S}\right)^2}, \tag{6.10a}$$

$$\mathfrak{A} = \frac{1}{\tau_{\mathrm{p}}}\left(\delta g - 2\frac{\beta}{S}\right), \tag{6.10b}$$

$$\mathfrak{B} = \frac{1}{\tau_{\mathrm{p}}}\delta g. \tag{6.10c}$$

It is obvious from (6.9) that the temporal evolution of the modal intensities possesses a time constant τ as given in (6.10a). This time constant decreases as δg increases and τ_{p} decreases. The dependence on δg is intuitively obvious since a higher modal discrimination leads to a faster time for the laser

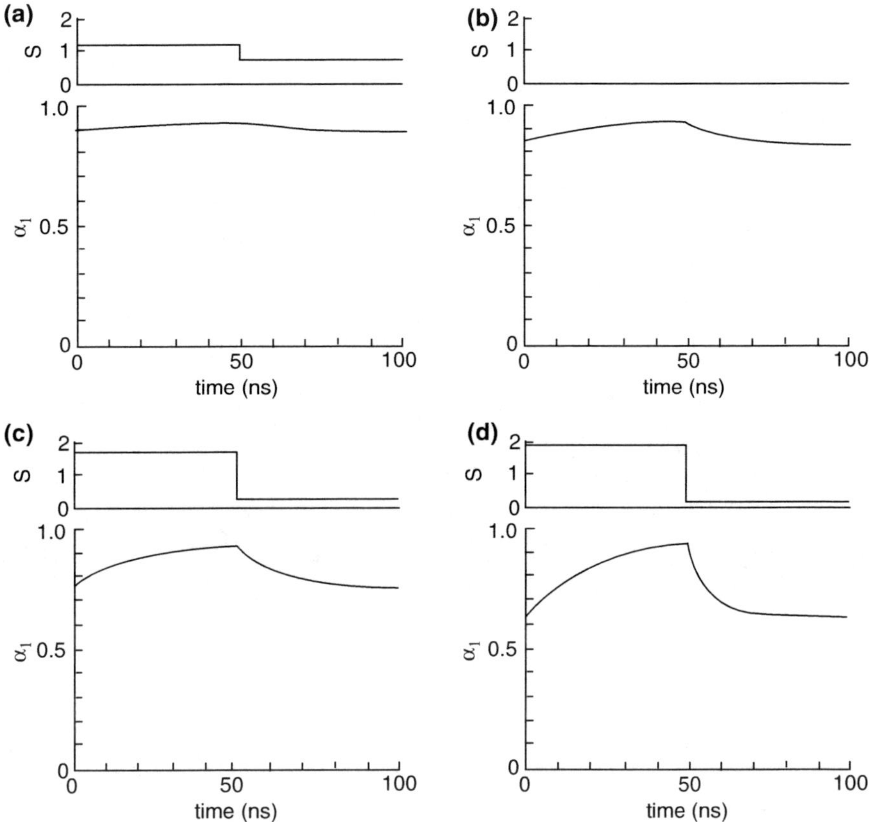

Fig. 6.4. Variation of the fraction of optical power in the dominant mode as a function of time in one cycle of a continuous square wave modulation at 10 MHz. The modulated waveform of the total photon density is shown above each plot. The optical modulation depths are 33%, 67%, 82%, and 95% in (**a**–**d**)

to equilibrate towards its steady state spectrum. A further examination of the solution indicates that at the limits of high and low photon densities, τ approaches the following:

$$\tau \sim \frac{2\tau_{\mathrm{p}}}{\delta g}, \quad S \text{ large}; \tag{6.11a}$$

$$\tau \sim \frac{S\tau_{\mathrm{p}}}{\beta}, \quad S \text{ small}. \tag{6.11b}$$

Thus, in pulse or square modulation where the bottom of the optical modulated waveform is fairly low, the time constant involved in the redistribution of spectral intensities is much shorter during turn-off than during turn-on. The time constant when the laser is turned on depends on the amount of mode discrimination δg whereas when the laser is turned off the time constant would

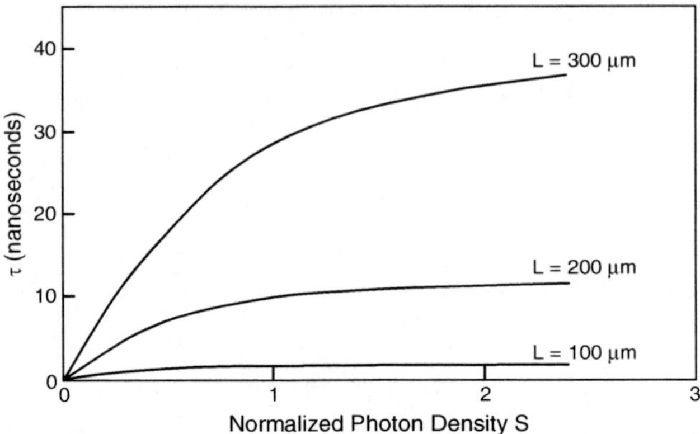

Fig. 6.5. Calculated time constant for spectral transients in a two-mode laser as a function of normalized total photon density. A normalized value of $S = 1$ corresponds roughly to an output power per facet of 1.5 mW. The values of the parameters used in the calculations are $1/\tau_{\mathrm{p}} = 1,500$, $b = 10^{-4}$ for a 300 µm device and is proportional to the square of cavity length, $\beta = 5 \times 10^{-5}$ for a 300 µm device and is inversely proportional to cavity length

depend on the optical power level at the off state. Figure 6.5 shows a plot of the time constant τ as a function of the total photon density S, for various cavity lengths. One can see from Fig. 6.5 and (6.10a) that, by reducing the length of the laser, one can reduce the time constant not only by an increased mode selectivity (δg) but also through a reduction in τ_{p}. Figure 6.4 shows plots of α_1, using (6.9), with the modulated waveform of the total photon densities shown in the top part of the figure. The modulation frequency is 10 MHz, and Fig. 6.4 shows cases with increasing modulation depth. These plots show that the time constant for equilibriating the spectrum is quite long (in the nanoseconds range) compared to the modulation period of the laser (when one looks at the total photon density). Thus, at high modulation frequencies (above $\sim$1 GHz), the spectral content does not have sufficient time to change from cycle to cycle, and the relative mode amplitudes are approximately constant in time. This is shown in Fig. 6.6, which has plots similar to Fig. 6.4, but at a higher modulation frequency of 300 MHz. The simple analytical results above are obtained by assuming that the total photon density takes the form of a square wave modulation, which intrinsically assumes that relaxation oscillation does not take place. However, in view of the fact that spectral transient processes are relatively slow ones, any rapid variation in the photon density during relaxation oscillation should not have significant effect on the solution, as can be seen from previous numerical results which showed that the spectral width rises and falls smoothly despite heavy oscillation in the optical output [44].

The above results indicate that the time constant for spectral dynamics is fairly long, on the order of 10 ns. This is longer than one actually observes,

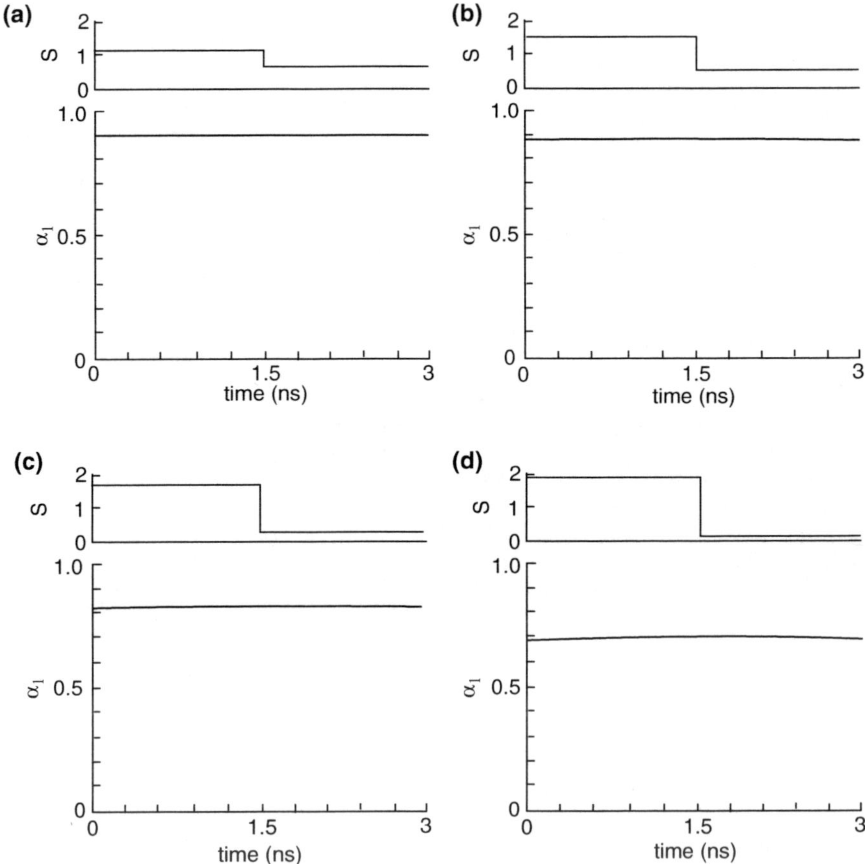

Fig. 6.6. Plots similar to Fig. 6.4 but a higher modulation frequency of 300 MHz. The optical modulation depths in (**a**–**d**) are identical to the corresponding plots in Fig. 6.4

and results from the fact that only two modes with almost identical gains are competing against each other. The analysis in the next section will show that, when many modes are taken into account, the time constant is considerably smaller – of the order of 0.5 ns. This is consistent with both experimental observations in GaAs lasers [46] and numerical results [44, 48].

6.5 Solution to the Many-Mode Problem

A meaningful description of the "purity" of the longitudinal mode spectrum of a semiconductor laser is the fraction of the total optical power contained in the dominant longitudinal mode α_0. which is described by (6.5)

$$\dot{\alpha}_0 = \frac{1}{\tau_{\mathrm{p}}} \left(\alpha_0 \sum_j (1 - g_j)\alpha_j - \frac{\alpha_0}{S} \sum_j \beta_j + \beta \right). \tag{6.12}$$

As mentioned before, an exact solution is not possible due to difficulties in evaluating the following time-dependent term in (6.12)

$$\sum_j (1 - g_j)\alpha_j(t). \tag{6.13}$$

However, from both numerical and experimental results previously reported, it seems reasonable to assume that the envelope of the multimode optical spectrum is Lorentzian in shape, whose width varies in time during modulation transients

$$\alpha_i(t) = \frac{\alpha_0(t)}{1 + c(t)i^2} \tag{6.14}$$

with the condition

$$\sum_i \alpha_i(t) = \sum_i \frac{\alpha_0}{1 + ci^2} = 1. \tag{6.15}$$

With this assumption, the summation (6.13) can be easily evaluated

$$\sum_i (1 - g_i)\alpha_i(t) = \alpha_0 \sum_i \left(\frac{bi^2}{1 + bi^2} \right) \left(\frac{1}{1 + ci^2} \right)$$
$$= \frac{b}{c - b} \left(\frac{\alpha_0 \pi}{\sqrt{b}} \coth \frac{\pi}{\sqrt{b}} - 1 \right), \tag{6.16}$$

where the relation in (6.15) has been used. The value of b is on the order of 10^{-4} whereas even in extreme multimode cases where there are ten or so lasing modes, the value of c is not much smaller than 10^{-1}. It is thus very reasonable to simplify (6.16) to

$$\sum_j (1 - g_j)\alpha_j(t) = \alpha_0 \pi \frac{\sqrt{b}}{c}, \tag{6.17}$$

where $\coth \pi x \to 1$ is used for $x \gtrsim 1$. An approximate analytical solution can be obtained for (6.14) and (6.15) whereby $c(t)$ can be expressed, with good accuracy, as an explicit function of $\alpha_0(t)$ (details in Sect. 6.5.1)

$$\frac{1}{c} = \frac{(1 + 2\alpha_0)(1 - \alpha_0)}{\pi^2 \alpha_0^2}. \tag{6.18}$$

Putting (6.17) and (6.18) into the time evolution equation (6.11), one has

$$\dot{\alpha}_0 = \frac{1}{\tau_{\mathrm{p}}} \left(\frac{\sqrt{b}}{\pi}(1 + 2\alpha_0)(1 - \alpha_0) - \alpha_0 \frac{\pi \beta}{S\sqrt{B}} + \frac{\beta}{S} \right). \tag{6.19}$$

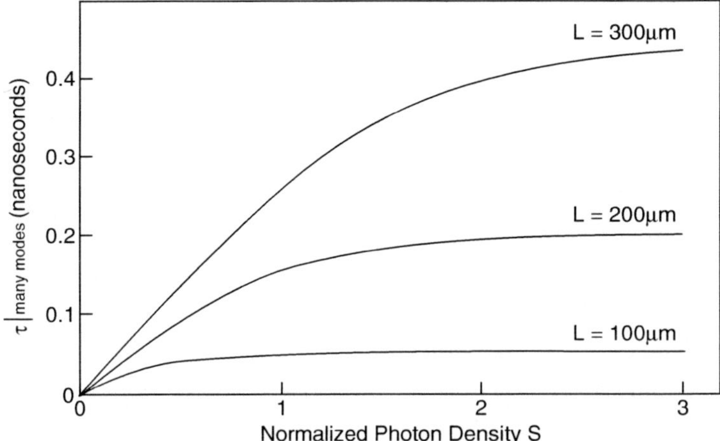

Fig. 6.7. Time constant for spectral transient when many longitudinal modes are taken into account. The values of the parameters are the same as that shown in the caption of Fig. 6.5

The form of this equation, $\dot{x} = Px^2 + Qx + R$ is similar to the time evolution equation in the two-mode case. The solution is thus similar in form to that discussed in the last section, (6.9). The corresponding time constant in this many-mode case can be evaluated from the coefficients in (6.19)

$$\frac{1}{\tau|_{\text{many mode}}} = \frac{1}{2\tau_{\text{p}}} \sqrt{\frac{9b}{\pi^2} + \frac{\pi^2\beta^2}{S^2 b} - \frac{2\beta}{S}}. \tag{6.20}$$

Figure 6.7 shows a plot of $\tau_{\text{many mode}}$ as a function of the total photon density S, for various cavity lengths. The similarity between these results and those in the two-mode case is apparent, except that the time scale involved here is considerably shorter. The reason is that there are many modes far away from the gain line center which take part in the transient process, as compared to the two-mode case where both of the modes are assumed to reside very closely to the line center. As a result, while the analysis in this section provides a fairly accurate description of the multimode spectral transient, it does not really add new physics or interpretation that could not be obtained in the two-mode solution. The time constants of about 0.5 ns for regular 300 μm GaAs lasers, as depicted in Fig. 6.7, agree well with experimental observations [46] and numerical computational results [44,57] for GaAs lasers. It also agrees well with recent experimental observations that coherent radiation can be obtained from a pulse-operated GaAs laser within 1 ns after the onset of the optical pulse [58]. On the other hand, experimental observations in quaternary lasers indicated a longer time constant of ∼5 ns. The reason for these observational differences is not clear. The general features of experimental observations, however, are in accord with the above theoretical results, where the time required for achieving the spectral steady state can be substantially reduced by increasing the amount of modal selectivity (i.e., increasing b), and where one can maintain an essentially single-mode spectrum as long as one maintains the laser above threshold [47].

6.5.1 An Approximate Analytic Solution of $\alpha_0 \sum_i \frac{1}{1+ci^2} = 1$

The above relation arose earlier in Sect. 6.5 relates the time evolution of the width of the (Lorentzian) spectral envelope, as measured by the quantity $c(t)$, in (6.14) to the time-varying fractional of optical power in the dominant mode $\alpha_0(t)$. A simple analytic solution expressing $c(t)$ as a function of $\alpha_0(t)$ was needed to further proceed with the analysis. This section provides such a solution. Using the relation

$$\sum_i \frac{1}{1+ci^2} = \frac{\pi}{\sqrt{c}} \coth \frac{\pi}{\sqrt{c}} \tag{6.21}$$

one arrives at a transcendental equation of the form

$$\alpha_0 x \coth x = 1, \quad \text{where } x = \frac{\pi}{\sqrt{c}} \tag{6.22}$$

and

$$0 < \alpha_0 < 1. \tag{6.23}$$

Consider the asymptotic behaviors as $\alpha_0 \to 0$ and $\alpha_0 \to 1$. One expects, from physical considerations, that the spectral envelope must be very wide if α_0 is very small, and therefore $c \to 0$ and $x \gg 1$, which justifies the approximation $\coth x = 1$, leading to

$$x = \frac{1}{\alpha_0}, \quad \alpha_0 \to 0. \tag{6.24}$$

On the other hand, as $\alpha_0 \to 1$, almost all the power is contained in the dominant mode and consequently the spectral envelope width should be very small: $c \to \infty$ and $x \to 0$. In this case one can expand $\coth x$ as

$$\coth x = \frac{1}{x} + \frac{x}{3} + \cdots \tag{6.25}$$

resulting in

$$x^2 = 3(1 - \alpha_0), \quad \alpha_0 \to 1. \tag{6.26}$$

Thus, asymptotically

$$x^2 = \begin{cases} (1/\alpha_0)^2 & \alpha_0 \to 0, \\ 3(1 - \alpha_0) & \alpha_0 \to 1. \end{cases} \tag{6.27}$$

In principle, it is possible to construct a solution for x^2, to an arbitrary degree of accuracy, with a rational function of α_0 which satisfies the asymptotic conditions (6.27). The simplest of such rational function is

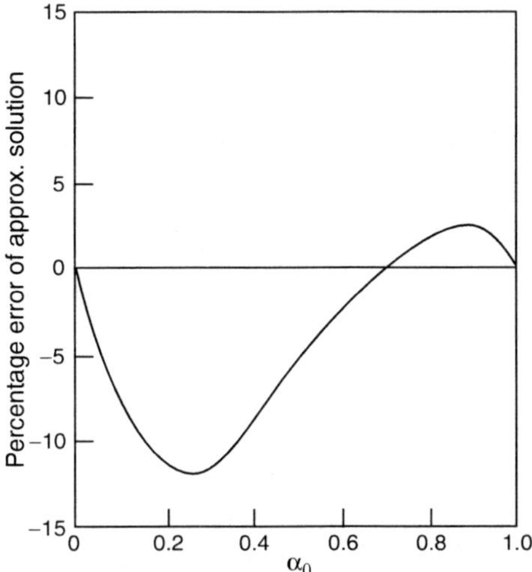

Fig. 6.8. The error percentage of the approximate solution (6.28) compared to the actual numerical solution, for various values of α_0

$$x^2 = \frac{(1 + 2\alpha_0)(1 - \alpha_0)}{\alpha_0^2}. \tag{6.28}$$

This solution, though simple, is remarkably accurate over range $0 < \alpha_0 < 1$. Figure 6.8 shows the error percentage of (6.28) compared to the true solution. The maximum error is about 12%. Replacing x by $\pi/\sqrt{c}$ in (6.28) yields the desired relation (6.18).

6.6 Lasing Spectrum Under CW High Frequency Microwave Modulation

In this section a quantitative comparison of the experimental results in Sect. 6.2 with the theoretical treatment of the last two sections will be performed. It is clear from the above analysis that under very high frequency continuous microwave modulation there is no significant *time variation* in the spectral envelope shape of the longitudinal modes, as evident from a *lack of time variation* of the *fractional* power content of the dominant mode (Fig. 6.6). Under this condition, the fraction of power in the dominant longitudinal mode α_0 can be deduced from the basic time evolution equation (6.5). Assume that the optical output power (and hence the total photon density) can be represented by

$$S(t) = S_0 + S_1 \cos \omega t. \tag{6.29}$$

Assuming that α_0 is constant in time, taking a time average (defined as $\langle \rangle = (1/T) \int_0^T dt$ where T = period of modulation) on both sides of (6.5) to give

$$\alpha_i \sum_j (g_i - g_j)\alpha_j - \alpha_i \sum_j \beta_j \left\langle \frac{1}{S(t)} \right\rangle + \beta_i \left\langle \frac{1}{S(t)} \right\rangle = 0. \tag{6.30}$$

The solution of (6.30) is the steady state lasing spectrum of a laser operating CW at a photon density of S_0', where

$$\frac{1}{S_0'} = \left\langle \frac{1}{S(t)} \right\rangle = (S_0^2 - S_1^2)^{-1/2}. \tag{6.31}$$

The optical modulation depth η, previously defined as the ratio of the amplitude to the peak of the optical modulated waveform, is

$$\eta = \frac{2S_1}{S_0 + S_1}. \tag{6.32}$$

Thus, in terms of modulation depth, the apparent dc power S_0' is

$$S_0' = S_0 \frac{2\sqrt{1 - \eta}}{2 - \eta}. \tag{6.33}$$

So, when a laser is biased at a certain optical power and being modulated at high frequencies with an optical modulation depth of η, the time-averaged lasing spectrum is equivalent to that of the laser operating CW (without modulation) at a reduced power level of S_0' as given in (6.32). Figure 6.9 shows a plot of the apparent reduction factor S_0'/S_0 vs. η. The results show that high-frequency modulation has little effect on the lasing spectrum unless

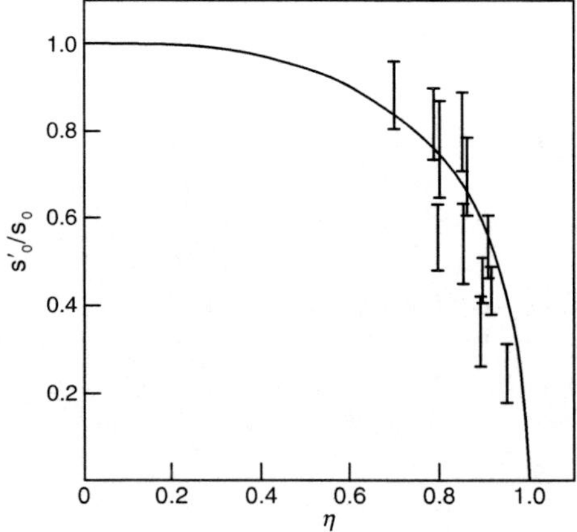

Fig. 6.9. A plot of S_0'/S_0 vs. η. S_0' is the photon density corresponding to a bias level, at which the laser would emit a longitudinal mode spectrum similar to that when the actual bias level is S_0 and the laser is modulated at high frequencies. The *vertical bars* are derived from experimental observations of the lasers shown in Figs. 6.1 and 6.2, and a few others

the optical modulation depth exceeds $\sim 80\%$. The points shown on the same plot are obtained from the experimental results of the two lasers described in Sect. 6.2 and a few other lasers. The general agreement with the analysis is good.

If one further increases the microwave drive to the laser beyond the point where the optical modulation depth approaches 100% ($S_0 \rightarrow S_1$), the bottom of the optical waveform will clip. The photon density during clipping will, in practice, be very small but not exactly zero. It is easy to see from (6.31) that as soon as clipping occurs the quantity $\langle 1/S(t) \rangle$ becomes very large and consequently S_0' becomes very small. The spectrum would look like that of a laser below lasing threshold. This is consistent with experimental observations.

It should be noted that the above result indicate that comparing the spectral purity of two lasers under the same optical modulation depth is not always fair-one has to consider the dc bias levels of these two lasers as well. Obviously, if one laser is biased way above threshold, so that even after being reduced by the factor shown in Fig. 6.9 the apparent bias level is still substantially above threshold, single-mode oscillation can be maintained up to very large optical modulation depths.

The analysis presented above is based on a strictly homogeneously broadened gain system and therefore does not take into account mode hopping and spectral gain suppression [52]. Spectral gain suppression is manifested as a decrease in the actual amplitude of the nondominating longitudinal modes as the optical power is increased, and is usually observed only at high optical power levels. This phenomena has been explained by nonlinear optical properties of the semiconductor material [59] and actually aids the laser in maintaining a single-mode spectrum under high-frequency modulation.

6.7 Dynamic Wavelength "Chirping" Under Direct Modulation

Sections 6.3–6.6 discuss excitation of multiple longitudinal modes under transient switching (Sects. 6.3–6.6) and cw microwave modulation (Sect. 6.7).

These multimode behaviors have obvious implications in fiber transmission in terms of deleterious effects to the signal due to dispersion of the fiber, even at the dispersion minimum of $1.3\,\mu\mathrm{m}$, since the wavelength separation between longitudinal modes are quite far apart for a typical laser diode. The effects are not insubstantial particularly for high frequency signals, even at the fiber dispersion minimum of $1.3\,\mu\mathrm{m}$. It is obvious from the discussions in Sects. 6.3–6.7 that a "single-wavelength" laser which remains single-wavelength even under transient switching or high frequency microwave modulation is *essential* for the possibility of fiber transmission over any reasonable distances. The results of Sects. 6.3–6.7 indicate that one single key parameter controls this behavior of the laser – namely the gain selectivity (difference) between the dominant mode and the neighboring modes – the factor δg in (6.8). For applications in telecommunications this issue has been solved by employing a highly wavelength selective structure, such as a grating, into the laser cavity, resulting in what is known as a Distributed Feedback Laser (DFB). Whereas mode rejection ratios (ratio of power in dominant mode to that of the next highest one) of $\sim$10–100 would have been considered excellent in Fabry–Perot lasers with cleaved mirror facets, and which relies on the slight difference in the intrinsic gain between longitudinal modes of the laser medium to produce mode selectivity. DFB lasers routinely exhibit mode rejection ratios in the thousands or higher, rendering moot the issue of multimode lasing and associated problems due to fiber dispersion. Despite the added complexity in fabrication of DFB lasers, their productions have now been mastered and volume productions is now a matter of routine from a number of vendors.

The issue which remains is that, while the lasing spectrum can be maintained to a single lasing mode under high speed modulation. It has been observed that the lasing wavelength (frequency) of that lasing mode can "chirp" under modulation. This phenomenon has its origin in the variation of refractive index of the semiconductor medium with electron density. It is obvious from previous considerations of the laser rate equations (1.19), (1.20) that under dynamic modulation situation the election density inside the laser medium does fluctuate along with the photon density. In fact, a definitive relationship between the time variations of the electron and photon density can be derived [60], with the result given in (7.6) and (15.24), where $\Phi(t)$ is the fluctuation of the phase of the electric field of the optical wave, $P(t)$ is the time varying photon density inside the laser medium; a time variation of the optical phase represents an (optical) frequency $(\sim \frac{\mathrm{d}\Phi(t)}{\mathrm{d}t})$, i.e., a wavelength dither, better known as "chirp" in the laser output wavelength. Note from (7.6) that the *entire* electric field, *including phase* of the optical output from the laser is known *deterministically* given the optical intensity waveform alone. It is most convenient that given any time-varying modulating current, the optical intensity waveform can be computed from the standard rate equations (1.19), (1.20), and then the wavelength (frequency) chirp can be computed from (7.6) accordingly. Armed with these results it is straight forward to compute the output optical waveform from a fiber link, given the dispersion and attenuation parameters of the fiber. These type of link simulations are now routinely done in the industry [61].

6.8 Summary and Conclusions

This chapter examines the dynamic longitudinal mode behavior of a laser diode under high speed modulation. Experimental observations of the lasing spectrum of a single-mode semiconductor laser under continuous microwave modulation show that the lasing spectrum is apparently locked to a single-longitudinal mode for optical modulation depths up to $\sim$80%, beyond which the lasing spectrum breaks into multimode oscillation. The width of the envelope of the multimode spectrum increases very rapidly with further increase in modulation depth. These results are satisfactorily explained by a theoretical treatment which gives simple analytic results for the time evolution of the individual longitudinal modes. It also yields considerable insight into spectral dynamics, and enables one to deduce the lasing spectrum of a laser under high-frequency modulation just by observing its CW lasing spectrum at various output powers. The results can also be used to deduce the amount of spectral envelope broadening under single or pseudorandom pulse modulation.

It is apparent from the results obtained in this chapter that single-mode oscillation can be maintained even under very high-speed modulation as long as one maintains the laser above lasing threshold *at all times*. A frequently quoted argument against doing so is that the added optical background can increase the shot noise level at the optical receiver. However, in an actual receiver system, the noise current is the sum of that due to shot noise and an effective noise current of the amplifier. The latter is of comparable amplitude to the former, and in many cases, is even the dominant of the two. The amount of added noise from reducing the optical modulation depth from 100%, to, say, 80%, may prove to be insignificant in most circumstances.

In addition to the excitation of multiple longitudinal modes when a laser diode is under current modulation another effect of significance is the so-called "*wavelength chirp*". This arises from a change in the refractive index of the semiconductor medium with fluctuation in electron density, which occurs when the current input into the laser diode is varied. This has been explained before when the laser rate equations were discussed in Chap. 2. It can be derived that [60] this effect produces a very profound result, that a definitive relation (15.25) exists between $E(t)$ and $P(t)$ where the former is the (*complex*) electric field output from the laser diode and the latter is the *power* output: $P(t) = |E(t)|^2$. It is straight forward to measure the time varying power output from the laser diode by using a photodiode, but it is not at all trivial to measure the time varying electric field (including optical phase). Equation (15.25) provides a convenient way to deduce the time varying E-field from a laser diode by measuring the time varying output power alone. For a derivation of this powerful relationship see [60].

Signal-Induced Noise in Fiber Links

7.1 Introduction

Common sources of noise in fiber-optic links include intrinsic intensity noise in the laser diode output arising from the discrete nature of electrons and photons (commonly known as "RIN," Relative Intensity Noise, and noises associated with the optical receiver. The latter, being relatively straight forward for an analog transmission system, at least in principle, will be briefly reviewed in Appendix B. This chapter is concerned with a quantitative evaluation of a scarcely discussed source of noise in subcarrier fiber transmission systems, namely, *"signal-induced* noise," (which only exists in the presence and in the spectral vicinity of a subcarrier signal). Specifically, one considers high frequency analog, single mode fiber-optic links using directly modulated multimode (Fabry–Perot) and single-frequency (DFB) lasers. The signal-to-noise ratio in a typical fiber-optic link is commonly evaluated by treating the various sources of noise, such as laser RIN, laser mode partition noise (for multimode lasers), shot and thermal noise of the receiver, etc., as *uncorrelated additive* quantities *independent* of the modulation signal. However, there are sources of noise which become prominent *only* in the presence of a subcarrier modulation signal. This chapter describes experimental and theoretical studies of this latter type of noise which arises from:

1. Mode-partitioning in (multi-longitudinal mode) Fabry–Perot lasers
2. Interferometrically converted phase-to-intensity noise in single-frequency DFB lasers

The former is greatly enhanced by fiber dispersion and the latter produced by optical retro-reflections along the fiber link such as multiple back reflections from imperfect connectors or splices. Even in the case where all connectors and splices are made to be perfect fundamental Rayleigh backscattering of the fiber glass material still serves as an ultimate cause of interferometrically converted phase-to-intensity noise. Both mode-partition noise in Fabry–Perot lasers coupled into a dispersive medium (fiber) as well as interferometrically converted phase-to-intensity noise in single mode lasers/fibers due to Rayleigh backscattering are well known. Both types of noise increase with fiber length, and so is the signal-induced noise which is created by these effects.

Historically, the development of high speed semiconductor lasers for microwave/analog applications took place with almost exclusive emphasis on intensity modulation speed, the reason being that, unlike fiber links in telecom or metropolitan networks this type of microwave/analog fiber links typically do not span any significant distances ($<$a few kilometers). As a result, $1.3\,\mu$m fiber at the dispersion minimum can be used in most of these applications thus obviating the concern for deleterious effects due to spectral impurity of the laser. To date, semiconductor lasers with the highest bandwidth has been demonstrated in Fabry–Perot (FP) lasers [62, 63], although single-frequency lasers also show exceptional performances [64]. The multimode lasing spectrum of FP lasers will be an issue in any wideband systems due to fiber dispersion, even for wavelengths near (but not exactly at) the fiber dispersion minimum of $1.3\,\mu$m. Another serious concern is the manifestation of mode-partition noise due to fiber dispersion, a subject studied previously in considerable detail [65]. The spectral content of this type of noise usually does not extend beyond a few tens of MHz and has been, generally (but erroneously), not considered harmful to high frequency microwave systems, hence they need not be considered for such applications. A similar type of low-frequency noise, the "mode-hopping" noise [65–67], which can manifest itself *without* the presence of fiber dispersion, was similarly considered not significant for narrowband microwave applications. However, it has been shown that in a directly modulated laser diode, the low frequency noise can be transposed to the spectral vicinity of the modulation subcarrier [68]. This is a source of signal-induced noise for Fabry–Perot lasers that can become quite serious at high frequencies and for long fiber links. These parameters will be quantified later. *It is therefore a mistake to not consider low frequency noise in lasers for high frequency applications.* Furthermore, it is equally misleading to measure the system noise level at high frequencies without any applied modulation to the laser, and then to calculate the anticipated S/N ratio based on these measurements, as if the signal and the noise are independent entities.

It should be noted that a similar transposition of low frequency noise onto the spectral vicinity of a high frequency modulation subcarrier also exists for systems with transmitters comprised of externally modulated diode-pumped YAG lasers. The low frequency noise from diode-pumped YAG lasers originates from relaxation oscillation (of the YAG laser), and from beating between longitudinal modes (of the YAG laser), the latter being *very* significant even with what would ordinarily be considered an excellent side-mode rejection. These noises typically do not extend beyond a few tens of megahertz. But given the fact that an external intensity modulator employed to modulate the microwave signal onto the optical carrier also acts as a mixer that multiplies the modulation signal with the intensity noises present on the optical beam, these low frequency noises will also appears in the spectral vicinity of the high frequency subcarrier signal (as a result of convolution of the spectra of the low frequency intensity noise and the modulation signal). Therefore these low frequency noises must be eliminated by optoelectronic feedback to the diode pump source of the diode-pumped YAG.

For single frequency diode lasers such as DFB lasers, neither mode-partition nor mode-hopping noises are present. The dominant source of low frequency noise arises from *double* back-reflections along the fiber which serves to convert the laser *phase noise* to *intensity noise* [65, 69, 70]. With the use of good optical connectors (Angled-Polished-Connectors, APC, for example), reflections along the fiber link can be minimized except for intrinsic Rayleigh backscattering along the length of the fiber [71]. The resulting intensity noise spectrum approximately resembles the (Lorentzian) laser lineshape centered at DC, with a typical linewidth of a few tens of megahertz. On applying high frequency direct modulation to the laser diode, this noise can be transposed to the spectral vicinity of the modulation subcarrier. In general, this effect is much less severe than that for FP lasers, provided proper optical connectors and splices are employed. In the remaining of this chapter these effects are described and parameters quantified.

7.2 Measurements

To illustrate the effects discussed above, Fig. 7.1 shows results for links consisting of (a) a $1.3\,\mu$m FP laser and (b) a $1.3\,\mu$m DFB laser, directly modulated at frequencies of 6.5 and 10 GHz, and propagating through distances of 1, 6, and 20 km of single-mode fiber. Both lasers are high-speed lasers resistively matched to 50 Ω input impedance, with a 3 dB modulation bandwidth well beyond 10 GHz, as shown in Fig. 7.1a. The same high-speed p-i-n photoreceiver with a 3 dB bandwidth of 12 GHz is used for all of the measurements. The optical emission spectra of the lasers are shown in Fig. 7.1b,c. The measurements are done with an input RF drive level into the lasers of +10 dBm. As a point of reference, the 1 dB compression level of the laser is +15 dBm. The optical input into the photoreceiver are adjusted to give 1 mA of dc photocurrent in all cases, except where noted in the caption. Angled polished optical connectors (APC) are used wherever a connection is required. The dispersion of the fiber used is estimated to be about 1 ps/(nm-km) at the wavelength of the lasers.

For the FP laser, low frequency mode-partition noise can be clearly observed on propagation through only a few kilometers of fibers, as shown in Fig. 7.2a,b. The highest of these noise levels correspond to RIN figures of $<-145\,\mathrm{dB\,Hz^{-1}}$ at the output of the laser,[1] $-132\,\mathrm{dB\,Hz^{-1}}$ after propagating through 6 km, and $-115\,\mathrm{dB\,Hz^{-1}}$ after 20 km. At higher frequencies, the noise drops back to the receiver noise limit of about $-145\,\mathrm{dB\,Hz^{-1}}$. In contrast, for the DFB laser the noise spectrum is at the receiver noise limit at

---

[1] This number is receiver noise limited.

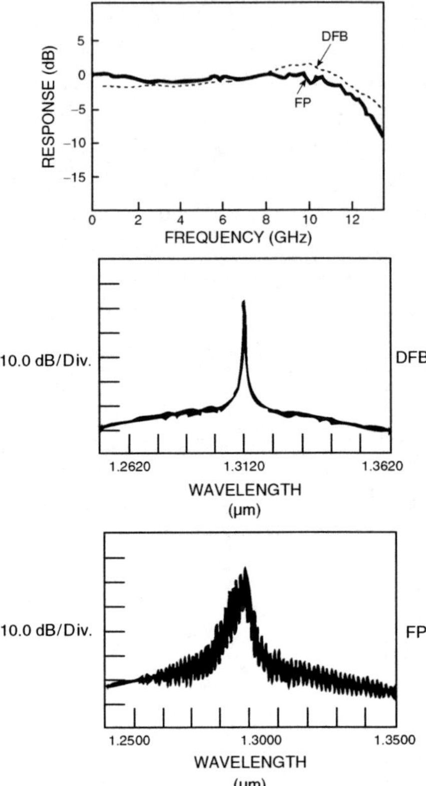

Fig. 7.1. (a) Direct modulation response of the high-speed FP and DFB lasers used in this test. Both lasers have an estimated relaxation oscillation frequency at slightly below 10 GHz, although the resonances of the both lasers are strongly damped. **(b)** and **(c)** lasing spectrum of the DFB and FP lasers, respectively. Note that the individual longitudinal modes of the FP laser are not resolved by the spectrometer in **(c)**

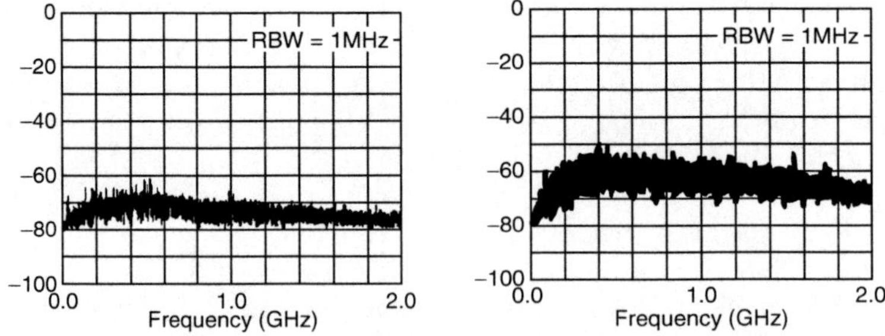

Fig. 7.2. Mode partition noise in FP lasers after transmission through **(a)** 6 km and **(b)** 20 km of single-mode fibers. Vertical scale in dBm, with DC photocurrent at the photodiode adjusted to 1 mA. The RF output from the photodiode is amplified by a 20 dB amplifier in these measurements

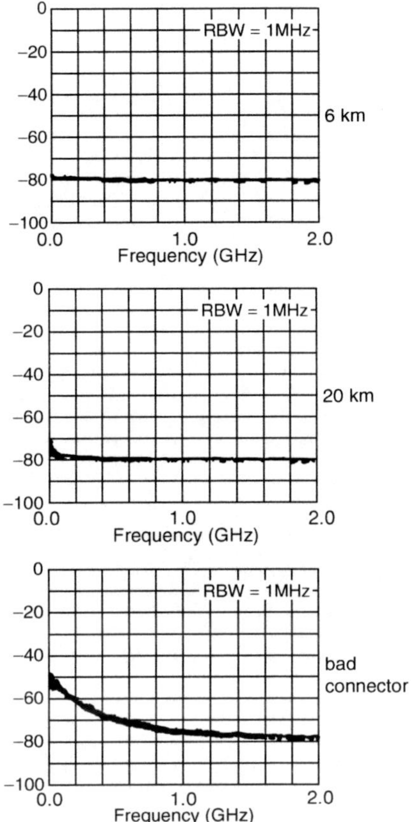

Fig. 7.3. Interferometric phase → intensity converted noise for DFB lasers due to double Rayleigh backscattering, for transmission through (**a**) 6 km and (**b**) 20 km of single-mode fibers, Angled Polished Connectors (APC) are used at all fiber interfaces. (**c**) Result of a bad splice in the fiber link. Vertical scale in dBm, dc photocurrent is adjusted to 1 mA in all cases. The RF output from the photodiode is amplified by a 20 dB amplifier in these measurements

6 km (Fig. 7.3a), and remains to be so after 20 km (Fig. 7.3b). Nevertheless, at 20 km one can already observe the interferometric noise emerging above the receiver noise level at low frequencies (Fig. 7.3b). This low level of interferometric noise is due to Rayleigh backscattering in a long fiber. The nature of the interferometric noise can be very easily observed when bad fiber splices or connectors exist in the link, as illustrate in Fig. 7.3c. A situation clearly to be avoided.

The kinds of low frequency noise described above are quite commonly observed in typical fiber links. In the following, it will be illustrated how the low frequency noise is transposed to the spectral vicinity of the modulation subcarrier upon application of a high frequency direct modulation to the laser.

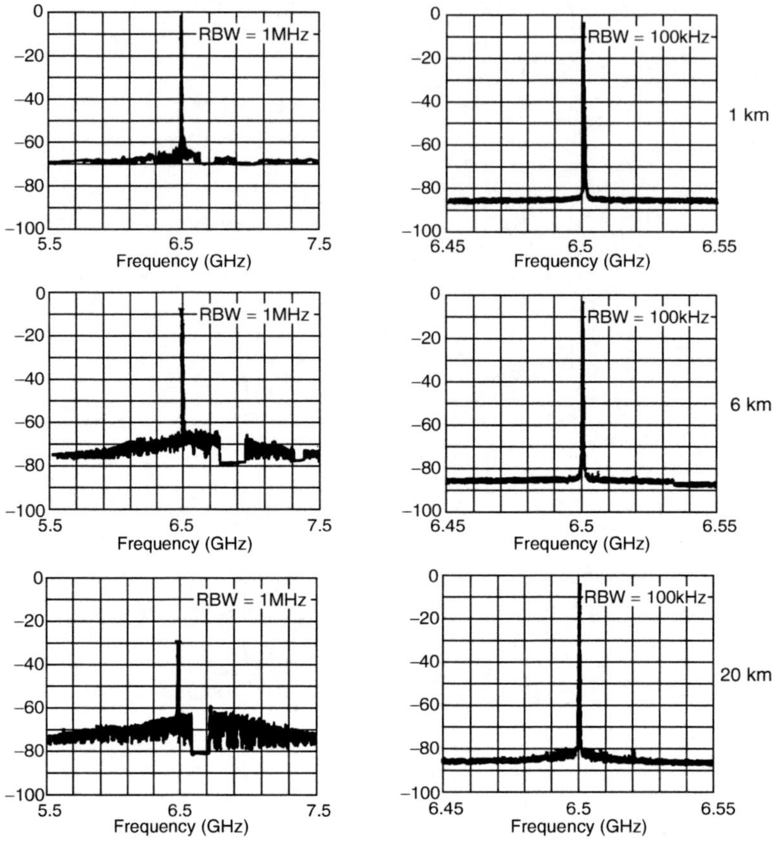

Fig. 7.4. RF noise spectra of photodiode output under an applied modulation signal at +10 dBm at 6.5 GHz. (**a**), (**c**), (**e**): FP laser, (**b**), (**d**), (**f**): DFB laser. Three cases are shown for each laser: transmission through 1 km (**a**) amd (**b**), 6 km (**c**) amd (**d**), and 20 km (**e**) amd (**f**) of single-mode fibers. The dc photocurrent is adjusted to be 1 mA in all measurements except (**e**), which is at 0.43 mA. Vertical scale in dBm. The RF output from the photodiode is amplified by a 20 dB amplifier in these measurements. The drop-outs' of the signal in (**c**) and (**e**) result from momentary removal of the optical input into the photo-receiver intentionally in order to establish a base line calibration for the background noise level of the measurement system

Figure 7.4 shows the result of applying a 6.5 GHz modulation subcarrier to the FP (a, c, e) and the DFB lasers (b, d, f), and observed after transmission through 1, 6, and 20 km of SMF. The blanks in the traces for FP laser in Fig. 7.4a–c were obtained with the RF drive to the laser disconnected, i.e., *without* noise transposition, in order to establish the background link noise level (dominated by intrinsic laser RIN). These plots illustrate that the low frequency noise and its transposition is very significant. The drop in the RF signal level at longer fiber lengths for the FP laser (Fig. 7.4c) is due to fiber dispersion and not attenuation (recall that *all* measurements, except where

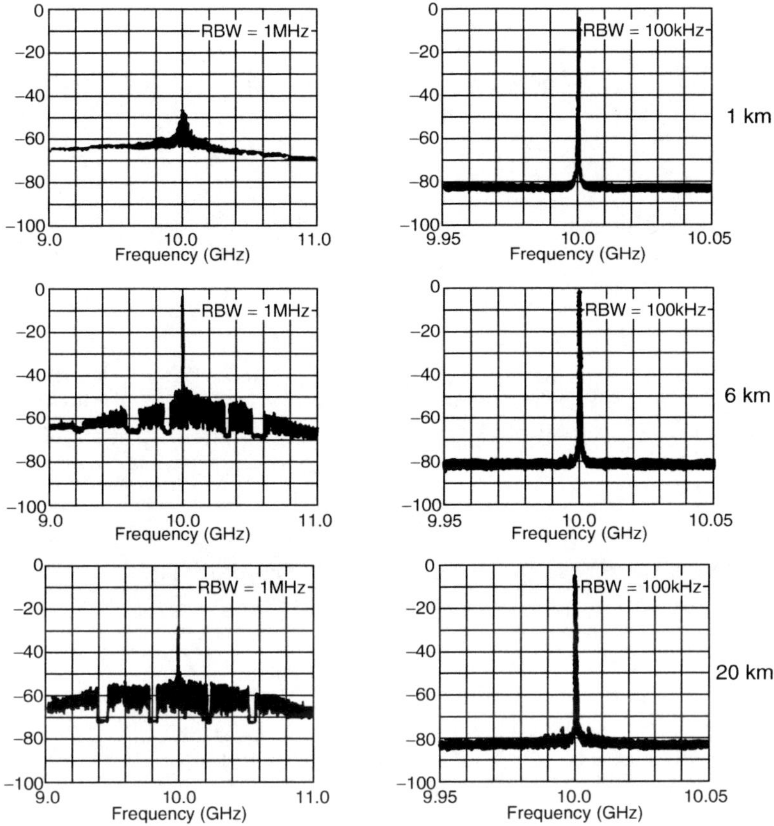

Fig. 7.5. Similar to Fig. 7.4 but at a modulation frequency of 10 GHz

noted, are done with 1 mA of photocurrent from the receiver). One should contrast the above results with those obtained with the DFB laser, Fig. 7.4d–f. In the latter cases only at 20 km does the transposed noise begin to emerge from the background, consistent with the low frequency noise observations of Fig. 7.3.

All of the above measurements were repeated with modulation applied directly to the laser at 10 GHz, as shown in Fig. 7.5a–f, note the even more striking difference between the FP laser (Fig. 7.5a–c) and the DFB laser (Fig. 7.5d–f). It is worthwhile to note that for a DFB laser, only a very slight degradation is observed even at 10 GHz and at 20 km, while the FP laser is all but unfunctional at these frequencies and distances. Note also from Figs. 7.4b and 7.5b that even at a relatively short distance of 6 km, the actual S/N ratio of the high-speed FP laser is, depending on the modulation frequency, between 10–20 dB worse than that predicted from a standard RIN measurement alone (without an applied modulation subcarrier and the concomitant noise transposition).

7.3 Analysis and Comparison with Measurements

Two principal sources of signal-induced noise which were illustrated in the above experiments are:

1. Transposed (from low frequencies) mode-partition noise in multimode lasers
2. Transposed interferometric noise (also from low frequencies) in single frequency lasers

Both mode-partition noise and interferometric noise, which are essentially low frequency noises, are already well known and have been studied extensively [65, 69–72]. The purpose of this section is to study the transposition of these noises to the spectral vicinity of a high frequency modulation subcarrier applied directly to the laser diode. The quantitative dependence of these transposed noises on signal frequency, amplitude and propagation length will be illustrated.

7.3.1 Mode Partition Noise and Noise-Transposition in Fiber Links Using Multimode Lasers

Even though a majority of linear fiber-optic links employ single mode fibers, there are situations which call for short lengths of legacy multimode fibers be used as patch chords to complete the connection. Issues related to noise behavior in multimode fibers should therefore be understood. This is the subject of this section. The well known deleterious effect in multimode fiber transmission is mode partition noise which is a result of competition between longitudinal modes in a multimode laser. The general properties of this noise are well known [67]. The modal noise characteristics can be derived by solving a set of electron and multimode photon rate equations driven by Langevin forces [65, 72]. This yields the complete noise spectra of each mode. If only the low frequency portion of the spectrum is of interest (anticipating that mode-partition noise does not extend to high frequencies), then one may neglect dynamic relaxation of the electron reservoir and obtain simpler results as in [72]. This latter approach does not suffice in the study here since one does need to consider the high frequency portion of the spectrum.

One can obtain closed form solutions to the mode partition noise problem in multimode lasers in the limit where only two modes exist, and where one mode dominates (i.e., a nearly single mode laser). This exercise yields insight into the nature of mode partition noise and its transposition to the spectral vicinity of a high frequency modulation subcarrier. Let S_1, S_2 be the *power* (photon density) in each of the two modes, with $S_1 \gg S_2$, then an approximate solution of the total relative intensity noise (RIN) of the optical output, after propagating through a length of dispersive fiber, is as follow: (see Sect. 7.4)

$$\mathrm{RIN} \times S^2 = 2R_{\mathrm{sp}}(S_1|A(\omega)|^2 + S_2|(1 - ke^{i\omega dL})B(\omega)|^2), \qquad (7.1)$$

where $S = S_1 + S_2$ is the total optical power, R_{sp} is the spontaneous emission rate into each mode, k is the relative coupling coefficient of mode 2 into the fiber (as compared to mode 1), d is the differential propagation delay of the two modes in the dispersive fiber, L is the length of the fiber, and

$$A(\omega) = \frac{i\omega + 1/\tau_{\mathrm{R}}}{(i\omega)^2 + i\omega\gamma_1 + \omega_{\mathrm{r}}^2}, \tag{7.2a}$$

$$B(\omega) = \frac{i\omega + 1/\tau_{\mathrm{R}}}{(i\omega)^2 + i\omega\gamma_2 + \delta\omega_{\mathrm{r}}^2}, \tag{7.2b}$$

where expressions for the corner frequencies ω_{r}, $\delta\omega_{\mathrm{r}}$, the damping constants γ_1, γ_2 and the effective lifetime τ_{r} can be found in the Sect. 7.4. As for the significance of these parameters, suffices to say that $\omega_{\mathrm{r}}/2\pi$ is the relaxation oscillation frequency in the direct modulation response of the laser, which is typically in the gigahertz range for high speed lasers, γ_1 is the damping constant for the direct modulation response which is approximately equal to $\omega_{\mathrm{r}}/2$ for a critically damped response typical of high speed lasers, while $\delta\omega_{\mathrm{r}}/2\pi$ is much smaller than $\omega_{\mathrm{r}}/2\pi$ – typically below $1\,\mathrm{GHz}$. The factor $B(\omega)$, which has a much lower corner frequency and a much higher dc value than $A(\omega)$, constitutes the mode partition noise. As evident from (7.1), the effect of this mode partition is visible only if the factor $1 - k\exp(i\omega dL) \neq 0$, which occurs if (a) the coupling of the two modes into the fiber (or the loss of the two modes) are not equal ($k \neq 1$), and (b) dispersion becomes significant ($\omega dL \gg 0$). It is also clear that a high mode rejection (defined by the ratio S_1/S_2) diminishes the effect of modal noise.

Figure 7.6 plots the theoretical RIN, (7.1) and (7.2) as a function of frequency for a highspeed laser with a $3\,\mathrm{dB}$ modulation bandwidth of $20\,\mathrm{GHz}$,

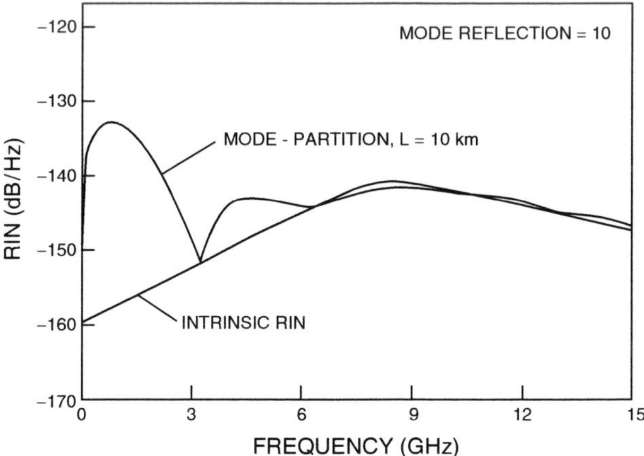

Fig. 7.6. Calculated mode-partition noise for a two-mode laser with mode rejection ratio of 10:1, on transmission through $10\,\mathrm{km}$ of fiber. Fiber dispersion was assumed to be $15\,\mathrm{ps\,km^{-1}}$ for the two modes under consideration

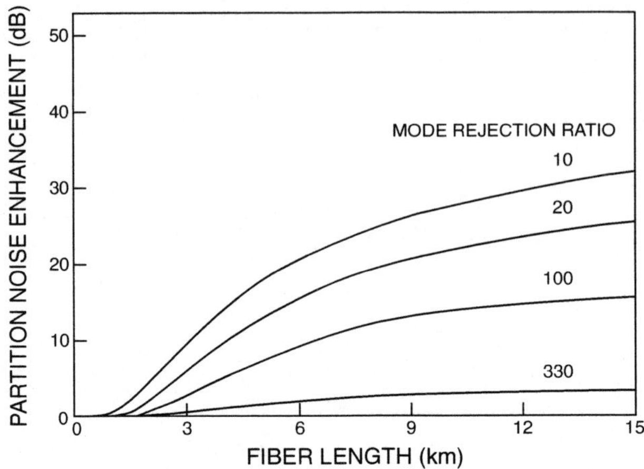

Fig. 7.7. Maximum enhancement in low frequency mode-partition noise as a function of fiber length, for various mode rejection ratios of the laser

before and after propagation through 10 km of single-mode optical fiber (SMF) with a dispersion of 15 ps km^{-1} between the two laser modes. The mode rejection ratio is assumed to be 10. The partial periodic structure of the RIN spectrum after propagation through 10 km of SMF is a result of the two-mode approximation. For lasers with multiple modes, the periodicity is largely suppressed due to the nonuniformity of fiber dispersion as a function of wavelength. Figure 7.7 plots the maximum noise enhancement due to fiber dispersion, as a function of fiber length, for different mode rejection ratios. This plot serves as a guideline for the required mode-rejection ratios required in order that modal noise becomes insignificant when propagated over a certain length of fiber. Mode rejection ratios of >100 are generally not achievable on a routine basis, particularly at 1.3 μm, without the use of mode-selective structures such as DFB.

The above RIN results apply for the case where no direct modulation is applied to the laser. When a high frequency modulation signal is applied, it has been shown that the low frequency intensity noise such as that generated by mode-partition is transposed to the spectral vicinity of the signal through intrinsic intermodulation effects in the laser diode [68]. This transposition depends on, among other things, the modulation signal frequency, and is described by a "noise transposition factor" $|T(\omega)|^2$, which is the amount of noise measured in the vicinity of the signal, when the optical modulation depth of the signal approaches 100%, compared to that of the low frequency noise without the applied modulation signal. The expression is given by [68]

$$T(\omega) = \frac{1}{2} \frac{(i\omega + \Gamma_1)(i\omega + \Gamma_2)}{(i\omega)^2 + i\omega\gamma_1 + \omega_r^2},$$ (7.3)

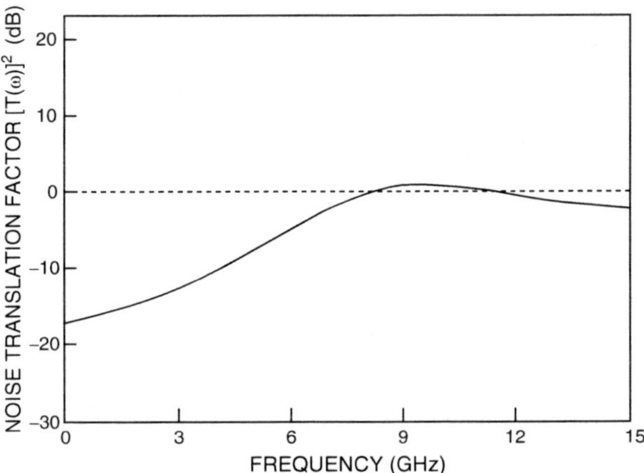

Fig. 7.8. Noise transposition factor for transposition of low frequency noise to high frequency, as a function of the applied modulation frequency. The optical modulation depth is assumed to be 1 in this plot

where $\Gamma_1 = \gamma_1 - 1/\tau_R$, $\Gamma_2 = \omega_r^2 \tau_p + 1/\tau_R$, τ_p is the photon lifetime. The noise transposition factor is plotted in Fig. 7.8, using the same laser parameters as in the previous plots.

Define $\mathrm{RIN}^{\mathrm{mod}}(\omega)$ as the measured RIN near the spectral vicinity of an applied modulation signal at frequency ω, at $\sim 100\%$ optical modulation depth. This quantity is a more meaningful description of the noise characteristic of the laser than the standard RIN when one is dealing with signal transmission with directly modulated lasers. It is given by

$$\mathrm{RIN}^{\mathrm{mod}}(\omega) = \mathrm{RIN}(\omega) + |T(\omega)|^2 \times \mathrm{RIN}(\omega = \omega_{\max}), \qquad (7.4)$$

where $\omega_{\max}$ is the frequency where the maximum in the low frequency mode-partition noise occurs (see Fig. 7.6). Using the results of Figs. 7.6 and 7.8, Fig. 7.9 plots $\mathrm{RIN}^{\mathrm{mod}}(\omega)$ vs. frequency, for different fiber lengths. For convenience in comparison, also included in figure are previous cases for $L = 10\,\mathrm{km}$ with and without the applied modulation, for a FP laser with a mode rejection ratio of 2 which is similar to the laser used in this measurements (Fig.7.10). It can be seen that the practical RIN in the spectral vicinity of the high frequency modulation subcarrier is enhanced to a value approximately identical to that of the low frequency RIN caused by mode-partition. This conclusion is supported by comparing the experimental results shown in Fig. 7.5b,c with that of Fig. 7.2. Included in Fig. 7.9 are data extracted from Figs. 7.4 and 7.5. The quantitative match is reasonably good considering the simplicity of the model.

In lasers where two or more longitudinal modes have nearly identical power, it has often been observed that a low frequency enhancement in RIN occurs at the laser output, even without propagating through any dispersive

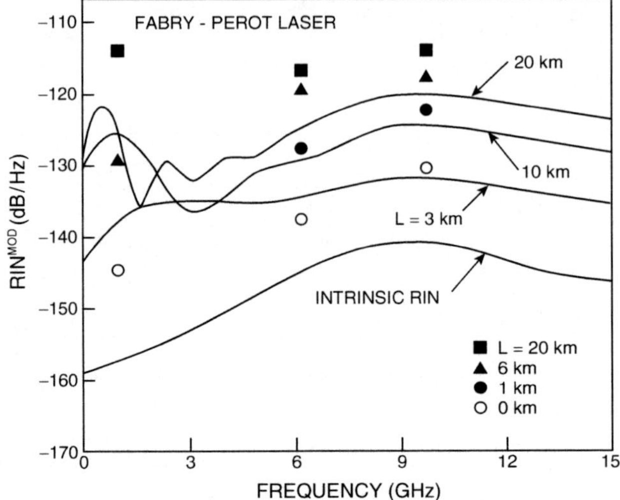

Fig. 7.9. Signal-induced RIN (RIN^{mod}), in FP lasers, as a function of modulation frequency, for different fiber transmission lengths. Data points are extracted from measurements in Figs. 7.4 and 7.5. The cases for 0 km (*blank circles*) also represent the cases where the applied modulation is turned off

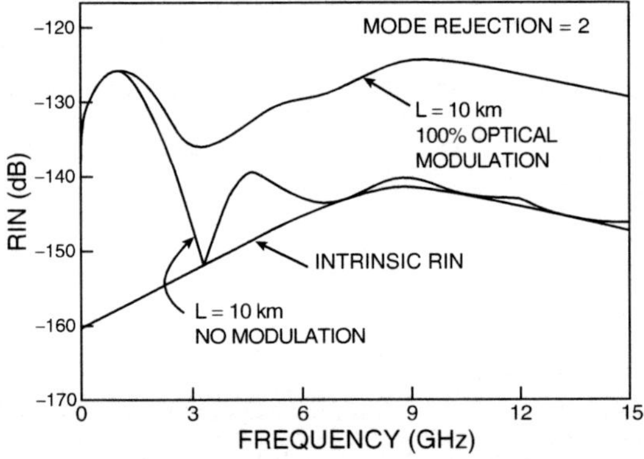

Fig. 7.10. A direct comparison of RIN without modulation and RIN with modulation for FP lasers, for 10 km transmission

fiber [6]. This effect is *not* explained by the standard small signal noise analysis using Langevin source as that outlined above, unless one includes a nonsymmetric cross gain compression between different longitudinal modes [73]. On the other hand, it is also possible that the enhanced noise is simply a large signal effect of mode competition: in principle, the damping effect of gain saturation, which is responsible for the suppression of RIN for the *total* power, is operative only in the small signal regime. It may well be that when mode-partition fluctuation in each mode is large, the delayed response in large signal situations [74] prevents instant compensation of power fluctuations between longitudinal modes, hence an enhanced noise at low frequencies. This is sometimes referred to as "mode-competition noise" or "mode-hopping noise," as distinct from "mode-partition noise." Regardless of the origin of these low frequency noise, the transposition effect is identical to that described above. The high frequency $\text{RIN}^{\text{mod}}(\omega)$ will again assume a value approximately equal to that of the low frequency RIN.

7.3.2 Transposed Interferometric Noise in Fiber Links Using Single Frequency Lasers

Interferometric noise is caused by conversion of the laser phase noise into intensity noise through interference between the laser output with a delay version of itself. This occurs in fiber links with pairs of interfaces where double reflection can occur, or in the absence of reflective interfaces, Rayleigh backscattering is the ultimate cause of such reflections [71]. In the absence of any applied modulation to the laser, this noise takes the form of a Lorentzian function, which is the result of an autocorrelation of the optical *field spectrum* transposed down to DC. The spectral width of this noise is therefore approximately that of the laser linewidth, in the tens of megahertz range, whereas the intensity is proportional to the power reflectivity p of the reflectors responsible for the interferometric phase $\rightarrow$ intensity noise conversion [71, 75]. For Rayleigh backscattering, this reflectivity is proportional to the length of the fiber for relatively short lengths of fibers, reaching a saturated value equal to the inverse of the attenuation coefficient for long fibers [71, 76].

The behavior of interferometric noise when the laser is directly modulated has been analyzed in the context of reduction of low frequency interferometric noise by an applied modulation at a high frequency [75, 76]. The nature of this reduction is that the noise energy at low frequency is transposed to the spectral vicinity of the harmonics of the applied modulation signal due to the large phase modulation associated with direct modulation of laser diodes [75]. This is desirable only when the applied high frequency modulation is simply used as a "dither" while the low frequency portion of the spectrum is used for transmission of baseband signal (i.e., the "information-bearing" signal). If the information is carried by the high frequency modulation itself, as in many microwave systems, then the transposed noise centered at the first harmonic of the applied signal is the undesirable "signal induced noise" which was the subject of previous sections. Following an approach similar to that in [75], if one assumes that the laser intensity is approximately given by

$$P(t) = P_0 \left(1 + \beta \cos \omega t\right), \tag{7.5}$$

where β is the optical modulation depth and P_0 is the average optical power, then the associated phase modulation is given by [65]

$$\phi = \frac{\alpha}{2} \left(\frac{\mathrm{d}}{\mathrm{d}t} (\ln P(t)) + \gamma_1 P(t)/P_0 \right), \tag{7.6}$$

where α is the linewidth enhancement factor, γ_1 is the damping constant in the direct modulation response of the laser at high operating power, as given in (7.16). This damping constant is related to fundamental laser parameters and is dominated by gain compression. For critically damped response common in high speed lasers, $\gamma_1 \sim \omega_r/2$ where ω_r is the relaxation oscillation frequency. Integrating (7.6) and neglecting higher harmonics in ϕ, one obtains

$$\phi(t) = \frac{\alpha a}{2}\sqrt{1+(\gamma_1/\omega)^2}\cos(\omega t + \psi_0),\tag{7.7}$$

where $\psi_0 = \arctan(\Gamma_1/\omega)$. The electric field from the laser is $E(t) = \sqrt{P(t)}\, E^{i\phi(t)}$.

Assume the laser field is twice reflected from a pair of reflectors with power reflectivity ρ separated by a distance $\tau\nu$ where ν is the group velocity in the fiber. The autocorrelation of the noise current arising from the field interfering with the twice-reflected version is [75]

$$(i_N(t)i_N(t+\delta\tau)) \equiv p_N(t,\delta\tau)\tag{7.8}$$

$$= 2P^2P_0^2 \cdot \sqrt{P(t)P(t-\tau)P(t+\delta\tau)P(t+\delta\tau-\tau)}R_-(\delta\tau)$$

$$\times \left[\cos 4\alpha\sin\left(\frac{\omega\tau}{2}\right)\sin\left(\frac{\omega\delta\tau}{2}\right)\cos\left(\omega t - \frac{\omega\tau}{2} + \frac{\omega\delta\tau}{2}\right)\right],$$

where R is the responsivity of the photodetector which is assumed to be 1 from now on, $\alpha' = \alpha/2\sqrt{1+(\gamma_1/\omega)^2}$, and $R_-(\delta\tau)$ is the autocorrelation of the laser field spectrum which constitutes the interferometric noise. After time-averaging,

$$R_N(\delta\tau) = 2\rho P_0^2 R(\delta\tau)$$

$$\cdot \left(1 + \frac{\alpha^2}{2}\cos\omega\tau + \frac{\alpha^2}{2}(1+\cos\omega\tau)\cos\omega\delta\tau + \cdots\right)$$

$$\times J_0\left(4\alpha'\alpha\sin\frac{\omega\tau}{2}\sin\frac{\omega\delta\tau}{2}\right).\tag{7.9}$$

Expanding the Bessel function J_0 in Fourier series,

$$R_N(\delta\tau) = 2\rho P_0^2 R_-(\delta\tau)[\Xi_1 + \Xi_2\cos(\omega\delta\tau) + (\dots)\cos(2\omega\delta\tau) + \cdots],\tag{7.10}$$

where

$$\Xi_1 = \left(1 - \alpha^2\sin^2\left(\frac{\omega\tau}{2}\right)\right) J_0^2\left(2\alpha'\alpha\sin\left(\frac{\omega\tau}{2}\right)\right)$$

$$+ \frac{\alpha^2}{2}\cos^2\left(\frac{\omega\tau}{2}\right) J_1^2\left(2\alpha'\alpha\sin\left(\frac{\omega\tau}{2}\right)\right),$$

$$\Xi_2 = 2\left(1 - \alpha^2\sin^2\left(\frac{\omega\tau}{2}\right)\right) J_1^2\left(2\alpha'\alpha\sin\left(\frac{\omega\tau}{2}\right)\right)$$

$$+ \frac{\alpha^2}{2}\cos^2\left(\frac{\omega\tau}{2}\right) J_0^2\left(\alpha'\alpha\sin\left(\frac{\omega\tau}{2}\right)\right).\tag{7.11}$$

The term involving Ξ_1 is the remnant of the low frequency interferometric noise, while that involving Ξ_2 is the transposed interferometric noise center at the signal frequency ω.

The factor $\Xi_1 (< 1)$ is periodic in $\omega\tau$, and was previously called the "noise suppression factor" [75, 76], in reference to the benefiting effect of interferometric noise suppression in the base band by an applied high frequency modulation. In the event where the applied modulation consists not just of a single tone but is of finite bandwidth, or that the locations of the reflectors are randomly distributed as in the case Rayleigh backscattering, the periodicity is removed [75, 76]. To evaluate these cases exactly, one needs to involve the statistics of these random distributions [75, 76]. However, the result is not very different from that obtained simply by averaging Ξ_1 over $\omega\tau\epsilon[0, 2\pi]$ [75].

The factor Ξ_2 is referred here as the "noise transposition factor" for interferometric noise, and is also periodic in $\omega\tau$ as in Ξ_1. Without going through the complication of accounting for the statistics of Rayleigh backscattering [76], approximate results can be obtained by simply averaging Ξ_2 over $\omega\tau\epsilon[0, 2\pi]$. This averaged noise transposition factor $\bar{\Xi}_2$ is plotted in Fig. 7.11 as a function of optical modulation depth β, at various modulation frequency ω. In the limit of $\beta \to 1$, the lower value of $\bar{\Xi}_2$ at low frequencies is a result of a higher effective phase modulation index (α').

To obtain the effective RIN under modulation ($\mathrm{RIN}^{\mathrm{mod}}$) like those shown in Fig. 7.9, assume that the Fourier transform of $R_{(\delta t)}$, which is the interferometric noise, is a Lorentzian with linewidth Δ. The power spectral density

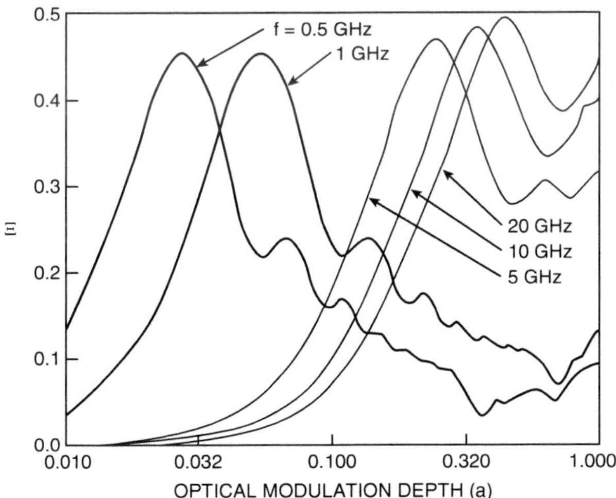

Fig. 7.11. Signal-induced noise transposition factor, Ξ, as a function of modulation frequency and optical modulation depth of the signal, for interferometric phase $\to$ intensity noise in DFB laser links

at DC is $1/\Delta$. The RIN of the transposed noise at the modulation signal frequency is therefore, from (7.10),

$$\mathrm{RIN}^{\mathrm{mod}} = p\frac{\Xi(\omega, \beta = 1)}{\Delta} + \mathrm{RIN}(\omega) + \mathrm{F.T.}(R_{(\delta\tau)}), \qquad (7.12)$$

where $\mathrm{RIN}(\omega)$ is the intrinsic intensity noise of the laser, and F.T. denotes Fourier transformation. Furthermore, one can use the previously derived [71] relation between the Rayleigh reflection coefficient ρ and the fiber length:

$$\rho^2 + W^2\left[\frac{L}{2\gamma} + \frac{1}{4\alpha^2}\left(1 - e^{-2\gamma L}\right)\right], \qquad (7.13)$$

where γ is the attenuation of the fiber per unit length, W is the "Rayleigh reflection coefficient per unit length," a constant that depends on fiber characteristics and typically takes on a value of 6×10^{-4} [71]. Using these results, one can plot in Fig. 7.12 the spectrum of $\mathrm{RIN}^{\mathrm{mod}}$ for fiber lengths of 10 and 20 km, assuming $\Delta = 20\,\mathrm{MHz}$ typical of a DFB laser. Observe that the signal induced noise in this case does not appreciably increase the high frequency RIN value even for fiber lengths up to 20 km. Also shown in Fig. 7.12 are data points extracted from measurement results, Figs. 7.4 and 7.5. There is a good match between theory and experiment, in particular note that at 20 km the high frequency signal induced noise is approximately 3 dB below that of the low frequency interferometric noise (compare Figs. 7.5f and 7.3b), which is the value given by $\bar{\Xi}_2$ shown in Fig. 7.11. One should also contrast these results for DFB lasers (Fig. 7.12) to that of FP lasers (Fig. 7.9). The superiority of the former is evident.

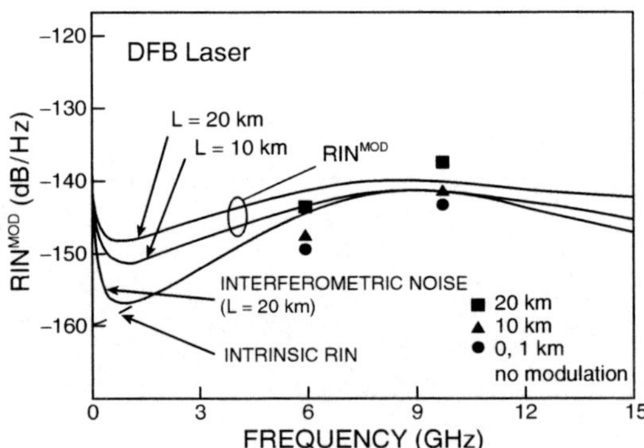

Fig. 7.12. Signal-induced RIN, $\mathrm{RIN}^{\mathrm{mod}}$, in DFB lasers, as a function of modulation frequency, for different fiber transmission lengths. Data points are extracted from measurements in Figs. 7.4 and 7.5. The cases where the applied modulation is turned off, and for 0 km transmission, are overlaid almost directly on top of the *solid circles*

7.4 Mode Partition Noise in an Almost Single Mode Laser

This section summarizes an approximate derivation of mode partition noise in a two-mode laser where one of the modes dominates. The approach is a standard one using multimode rate equations [65], and the results are simplified using a two mode approximation with one dominant similar to that in [73]. Let S_i be the photon density in the ith longitudinal mode. The multimode rate equations are [65]:

$$\frac{\mathrm{d}N}{\mathrm{d}t} = \frac{J}{ed} - \frac{N}{\tau_\mathrm{s}} - \nu \sum_i g_i(N)S_i + F_N(t), \tag{7.14a}$$

$$\frac{\mathrm{d}S_i}{\mathrm{d}t} = \nu \Gamma g_i(N)S_i - \frac{S_i}{\tau_\mathrm{p}} + R_\mathrm{sp} + F_{S_i}(t), \tag{7.14b}$$

where N is the carrier density, Γ is the optical confinement factor, J is the pump current density, d the thickness of the active region, τ_s is the recombination lifetime (radiative and non-radiative) of the carriers, τ_p is the photon lifetime, $g_i(N)$ is the optical gain of the ith mode as a function of the carrier density, expressed in cm^{-1}, ν is the group velocity, R_sp is the spontaneous emission rate into each mode, and e the electronic charge. $F_N(t)$ and $F_{S_i}(t)$ are Langevin noise sources driving the electrons and the modes; their correlation characteristics have been derived in detail [77–79]. The form of optical gain is assumed to be

$$g_i(N) = g'_{i0}(N - N_0)(1 - \nu \epsilon S_i), \tag{7.15}$$

where ϵ is the gain compression parameter, N_0 is the transparency electron density, and g'_{i0} assumes a parabolic gain profile near the gain peak. Note that cross-compression terms between different longitudinal modes were neglected. Its effects has been studied previously [73], and is shown to produce a low frequency "mode-hopping noise" in situations where two or more longitudinal modes have almost equal power. The noise spectra are obtained by a small signal solution of (7.14), using the proper Langevin correlation characteristics. In the case of a nearly single-mode laser ($S_1 \gg S_2$), one can obtain the noise spectra in closed (albeit approximate) form in an approach similar to that used in [73]:

$$s_1(\omega) = F_{s_1}(\omega)A(\omega) - F_{s_2}(\omega)B(\omega), \tag{7.16a}$$

$$s_2(\omega) = F_{s_2}(\omega)B(\omega), \tag{7.16b}$$

where $s_1(\omega)$, $s_2(\omega)$ are Fourier transforms of the small signal modal fluctuations, $A(\omega)$ and $B(\omega)$ are given by (7.2), with the following parameters:

$$\gamma_{1,2} = \frac{R_{\mathrm{sp}}}{S_{1,2}} + \frac{1}{\tau_{\mathrm{R}}} + \nu\epsilon S_{1,2}, \tag{7.16c}$$

$$\omega_{\mathrm{r}}^2 \sim g_1' S_1/\tau_{\mathrm{p}}, \tag{7.16d}$$

$$\delta\omega_{\mathrm{r}}^2 = \left[\frac{R_{\mathrm{sp}}}{S_2} + \nu\epsilon S_2\right]\frac{1}{\tau_{\mathrm{R}}}, \tag{7.16e}$$

τ_{R} is the effective carrier lifetime (stimulated and spontaneous), $F_{s_1}(\omega)$, $F_{s_2}(\omega)$ are Fourier transforms of the Langevin noise driving the respective modes, with the following correlation relations [79]:

$$\langle F_{s_i}(\omega)F_{s_j}^*(\omega)\rangle = 2R_{\mathrm{sp}}S_i\delta_{ij}. \tag{7.17}$$

It has been assumed, as is customarily the case, that the Langevin force driving the electron reservoir $(F_N(t))$ is negligible.

The small signal fluctuation of the total power after propagating through a dispersive fiber is

$$s(\omega) = s_1(\omega) + s_2(\omega)e^{i\omega dL}, \tag{7.18}$$

where d and L are the differential delay between the two modes per unit length and fiber length, respectively. The RIN is given by

$$\mathrm{RIN} \times (S_1 + S_2)^2 = \langle s(\omega)s^*(\omega)\rangle, \tag{7.19}$$

which can be evaluated using the correlation relation (7.17). The result is given in (7.1) in Sect. 7.3.1.

7.5 Conclusion

A quantitative comparison is made, both theoretically and experimentally, of signal-induced noise in a high frequency, single-mode fiber-optic link using directly modulated multi-mode (Fabry–Perot) and single-frequency (DFB) lasers. It is clear that the common practice of evaluating the signal-to-noise performance in a fiber-optic link, namely, treating the various sources of noise independently of the modulation signal, is quite inadequate in describing and predicting the link performance under real life situations. This type of signal-induced noise arises from mode-partitioning in Fabry–Perot lasers, and interferometric phase-to-intensity noise conversion for links using DFB lasers, the former induced by fiber dispersion and the latter by fiber reflection caused by Rayleigh backscattering (assuming no bad splices in the fiber link). Both of these effects increase with fiber length, and so does the signal-induced noise brought about by these effects. Both of these types of noise concentrate at low frequencies, so that a casual observation might lead to the conclusion that they are of no relevance to high frequency microwave systems. Experimental observations described above indicate that this is not the case even for narrow band transmission at high frequencies through moderate lengths of fiber, FP lasers are *unacceptable* not just from consideration of transmission bandwidth limitation due to fiber dispersion, but from the detrimental *effect of signal-induced noise* due to mode-partitioning. For example, degradation in S/N performance is already significant in transmission of a 6 GHz signal over only 1 km of single-mode fiber. However, with DFB lasers, there is no degradation of the S/N performance for transmission at 10 GHz even up to 20 km.

Part II

Direct Modulation of Semiconductor Lasers Beyond Relaxation Oscillation

8
Illustration of Resonant Modulation

Former chapters in Part I of this book discussed present understanding of direct modulation properties of laser diodes with particular emphasis on modulation speed. A quantity of major significance in the small-signal modulation regime is the $-3\,\mathrm{dB}$ modulation bandwidth, which is a direct measure of the rate at which (primarily baseband) information can be transmitted by intensity modulation of the laser. However, one can obtain a large modulation optical depth (in fact, pulse-like output) at repetition rates beyond the $-3\,\mathrm{dB}$ point by driving the laser with sufficient RF drive power to compensate for the drop-off in the modulation response of the laser. This technique is very useful in generating repetitive optical pulses from a laser diode at a high repetition rate, although the repetition rate itself has *no* significance in terms of information transmission capacity of the laser. A means to reduce the RF drive power required for modulating the laser to a large optical modulation depth at high repetition rate is the technique of "mode locking." The laser diode is coupled to an external optical cavity whose round-trip time corresponds to inverse of the modulation frequency applied to the laser diode. The modulation frequency in this scenario is limited to a very narrow range near the *"round-trip frequency"* (defined as inverse of the round trip time) of the external cavity. An example of this approach used a $\mathrm{LiNbO_3}$ directional coupler/modulator to produce optical modulation at $7.2\,\mathrm{GHz}$ [80]. Another example involved coupling the laser diode to an external fiber cavity [81] which produced optical modulation up to $10\,\mathrm{GHz}$. Chapter 4 describes experimental work which extended the small-signal $-3\,\mathrm{dB}$ direct modulation bandwidth of a *solitary* laser diode to $\sim\!12\,\mathrm{GHz}$ using a "window" buried heterostructure laser fabricated on semi-insulating substrate (BH on SI) [27].

This chapter describes results of modulation of this "window BH on SI" laser at frequencies beyond the $-3\,\mathrm{dB}$ point, in both the small signal and large signal regimes. It will be described below that lasers operating in this mode can be used as a narrowband signal transmitter at frequencies beyond the $-3\,\mathrm{dB}$ point (or relaxation oscillation frequency), with a reasonably flat response over a bandwidth of up to $\sim\!1\,\mathrm{GHz}$. The response of the original solitary laser at this frequency range is substantially lower than that in the baseband range (i.e., at frequencies < relaxation resonance) and consequently high power RF drivers are necessary to attain a sufficient optical modulation depth for communications purpose. It was found that a weak optical feedback from an external optical cavity can boost the response by a substantial amount over a broad frequency range around the round-trip frequency of the external cavity. A strong optical feedback produces a sharp spike in the response of the laser at the round-trip frequency of the external cavity (hereafter called "on-resonance"). Under this condition, picosecond optical pulses can be generated by applying a strong current modulation to the laser on resonance, which can be interpreted as *active mode locking* [82] of the longitudinal modes of the composite cavity formed by the coupling of the laser diode and the external cavity.

The laser used in this experiment was a GaAs/GaAlAs "window BH on SI" laser described in Chap. 4. The length of the laser is 300 μm, with an active region dimension of 2 μm×0.2 μm. The presence of the transparent window near the end facet alleviates the problem of catastrophic damage and enables the laser to operate at very high optical power densities. The tight optical and electrical confinement along the length of the laser cavity (except at the window region) enables maximum interaction between the photon and electrons to take place and results in a very high direct modulation bandwidth. The small-signal modulation bandwidth of this device biased at an optical output power of 10 mW is shown as the dark solid curve in Fig. 8.1. Here, the "small-signal" regime is loosely defined as that when the modulation depth of the optical output is ≲80%. The −3 dB bandwidth, as shown in Fig. 8.1,

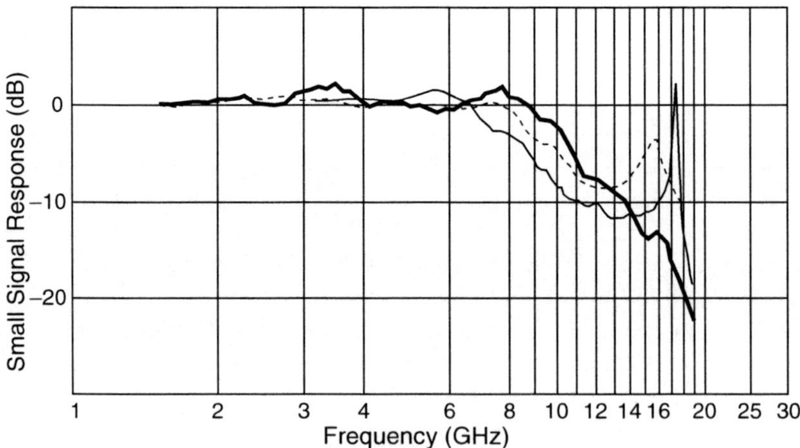

Fig. 8.1. Small-signal modulation response of a window BH on SI laser: (**a**) intrinsic laser response (*dark solid curve*); (**b**) weakly coupled to an external fiber cavity (*dotted curve*), and (**c**) with increased coupling (*light solid curve*). (From [83], ©1985 AIP. Reprinted with permission)

is 10.3 GHz. The response drops to $-10\,$dB at $\sim$13.5 GHz and to $-20\,$dB at 18 GHz. The fall-off in the modulation response is due to a combination of the intrinsic laser response and effects due to parasitic elements. A detailed examination of the modulation response characteristic, shows that in a 1 GHz band, centered at 16 GHz. The response is relatively flat over the 1-GHz band (to within $\pm 2\,$dB), and is within $\pm 1\,$dB over a 100-MHz bandwidth. It is thus possible to use this laser as an optical transmitter operating in a narrow bandwidth in the upper X-band range.

The intrinsic modulation response (i.e., that of the laser diode without the external fiber cavity) in Fig. 8.1 shows that at 16 GHz, the response is approximately 13 dB below the baseband value. (The small peak in the modulation response at around 16 GHz is probably due to electrical reflection arising from imperfect impedance matching of the laser.) It was found that this loss in modulation efficiency can be partially compensated for by coupling the laser to an external optical cavity of the appropriate length namely, that which corresponds to the "round-trip frequency." In this experiment the external cavity was composed of a short length (6.3 mm) of standard graded index multimode fiber of 50-µm core diameter [81, 84], with a high refractive index hemispherical lens attached to one end of the near end of the fiber to facilitate coupling. The far end of the fiber is cleaved but not metalized. The amount of optical feedback into the laser in this arrangement was estimated to be below 1%, and produces no observable reduction in lasing threshold or differential quantum efficiency. The feedback, however, induces a broad resonance in the frequency response at $\sim$16 GHz – the round-trip frequency of the fiber cavity – as shown by the dashed curve in Fig. 8.1. The full width of the resonance is about 1.5 GHz, measured at the upper and lower $-3\,$dB points. At the peak of the resonance the modulation efficiency is enhanced by $\sim$10 dB as compared to that of the laser without the fiber external cavity. The $-3\,$dB bandwidth of the resonance is approximately 1.5 GHz.

In a separate experiment the far end of a fiber was cleaved and butted to a gold mirror (with index-matching fluid in the small gap between the fiber facet and the gold mirror). This induced a very sharp resonance in the modulation response of the laser, as shown by the light solid curve in Fig. 8.1. When the laser is driven on resonance by a microwave source with an RF drive power of $-6\,$dBm, the optical output is not fully modulated and the laser is operating in the small-signal regime. As the microwave drive power is increased to $>10\,$dBm the optical modulation depth approaches unity and the optical waveform becomes pulse-like. The detailed characteristics of the optical pulses cannot be resolved by the photodiode, whose output appears to be sinusoidal since only the fundamental frequency (17.5 GHz) of the modulated laser light can be detected with reasonable efficiency. Figure 8.2 shows optical Second Harmonic Generation (SHG) autocorrelation traces of the laser output under two microwave drive power levels, at +4 dBm and at +14 dBm. The first trace (at +4 dBm drive) is sinusoidal in shape, implying that the optical waveform is also sinusoidal, and that the optical modulation depth is less than unity.

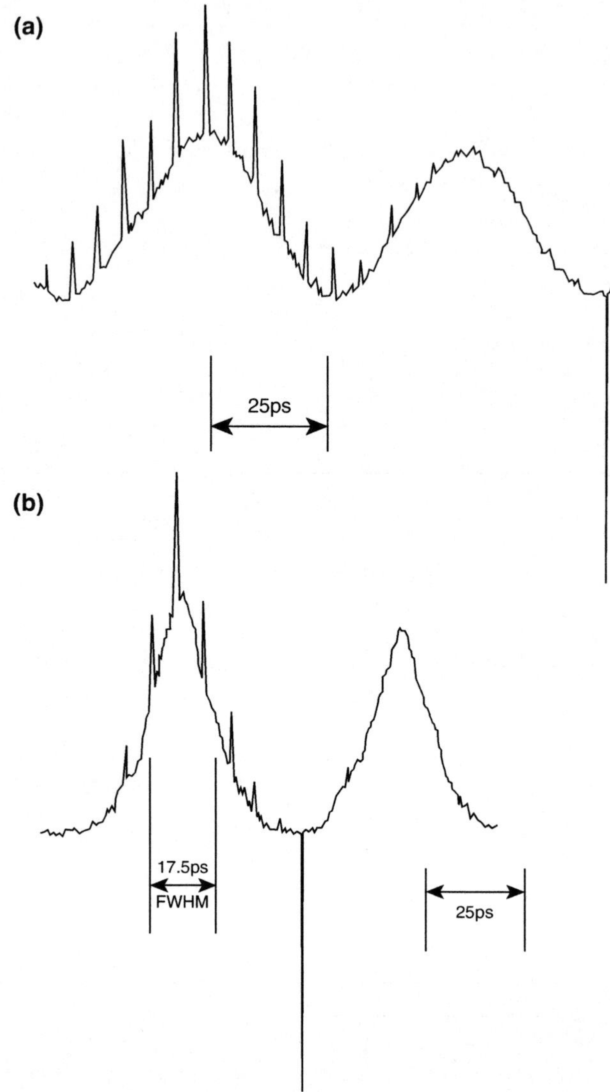

Fig. 8.2. Autocorrelation of the optical output of the window BH on SI laser coupled to an external fiber cavity under (**a**) 4-dBm microwave drive and (**b**) 14-dBm microwave drive at 17.5 GHz. (From [83], ©1985 AIP. Reprinted with permission)

The latter case (b) clearly indicates the pulse-like behavior of the optical output, with a full width at half-maximum (FWHM) width of 12.4 ps (inferred from the FWHM of the autocorrelation trace, assuming a Gaussian pulse shape). This, in effect, is active mode locking of a laser diode at a repetition rate of 17.5 GHz. The spectrum of the laser consists of a large number ($\sim$7) of longitudinal modes of the laser diode since there is no frequency selective element (such as an etalon) in the external cavity. The width of the individual mode is mainly determined by frequency chirping due to heavy carrier modulation and does not seem to correspond to the transformed value of the optical pulse width.

There is a subtle but important difference between short optical pulse generation by large signal modulation of a solitary laser diode and by active mode locking. In the former case, each optical pulse builds up from essentially spontaneous emission noise and therefore pulse to pulse coherence is very poor or non-existent. In the latter case, each pulse builds up (at least partially) from stimulated emission of the previous optical pulse returning from a round-trip tour of the external cavity, and hence successive optical pulses are coherent to each other. However, the autocorrelation traces of Fig. 8.2a,b show that pulse to pulse coherence is quite poor in the output of these very high rate actively mode-locked lasers. This is most likely due to:

1. The large amount of frequency chirping due to variations in the refractive index of the laser material at such high modulation frequencies [85, 86].
2. The relatively small feedback from the external cavity

The above experiments demonstrated that suitably constructed high speed laser diodes can be used as narrowband signal transmitters in the Ku band frequency range (12–20 GHz). The modulation efficiency can be increased considerably by a weak optical feedback to the laser diode. A stronger optical feedback enables one to *actively mode lock* the laser diode at a high repetition rate of up to 17.5 GHz, producing pulses ~12 ps long. Short optical pulse trains at very high repetition rates has been suggested for use as an optical frequency comb standard for locking the wavelengths of laser transmitters in a Dense Wavelength Division Multiplexed system. Furthermore, the above mode-locking experiment points to a possible means of modulation of an optical carrier by narrow-band microwave signals beyond the limit imposed by the classic relaxation oscillation limit. It will be shown in Chap. 9 that this concept can be extended to mm-wave frequencies.

Furthermore, experimental results to be described in Chap. 10 show that this modulation scheme does has sufficient analog fidelity for meaningful signal transmission in the mm-wave range. It is ironic that the scheme described above is easier to implement in practice in the mm-wave range than at lower frequencies, since the higher frequency range necessitates a shorter optical cavity to the extent that at frequencies of $\gtrsim$50–60 GHz *monolithic* laser devices can be used without the need for a cumbersome external optical cavity, which invariably complicates the task of reliable packaging.

9

Resonant Modulation of Monolithic Laser Diodes at mm-Wave Frequencies

Most millimeter wave systems ($\gtrsim 70\,\mathrm{GHz}$) operate in a relatively narrow bandwidth albeit at frequencies much above the presently attainable direct modulation bandwidth of laser diodes. In this chapter, it is shown that through the use of active mode locking technique [87] efficient direct optical modulation of semiconductor lasers at frequencies up to and beyond $100\,\mathrm{GHz}$ is fundamentally possible. In the present literature the term "mode-locking" is synonymous with short-pulse generation, in which many longitudinal modes are locked in phase. Here it is used in a more liberal sense to encompass effects resulting from modulation of a laser parameter at the inter-longitudinal modal spacing frequency, which is identical to the "round-trip frequency" defined before in Chap. 8, even when it results in the phase locking of only a small number of modes (2–3).

Previous efforts on passive and active mode-locking laser diodes, with [83, 89–91] or without [92] external cavities, have approached frequencies slightly below $20\,\mathrm{GHz}$. To ascertain the fundamental limit to which the highest frequency which mode-locking can take place, the active mode locking process is first analyzed using a self-consistent approach [93], as shown in Fig. 9.1a, in which the gain modulation is not treated as a prescribed entity as in standard analysis [94], but which interacts with the optical modulation resulting from it [93]. To begin, assume that the electron density varies sinusoidally in time with a frequency of Ω:

$$N = n_0 + 2n \cos \Omega t. \tag{9.1}$$

In anticipation of only a small number of participating modes at high frequency mode-locking. Three modes are included in the analysis with amplitudes A_0 and $A_{\pm 1}$. The mode-coupling equations are [94, 95]

$$A_0 \left(-\frac{1}{2\tau_{\mathrm{p}}} + \frac{n_0 G}{2} \right) = -\frac{nG\xi}{2}(A_1 + A_{-1}), \tag{9.2}$$

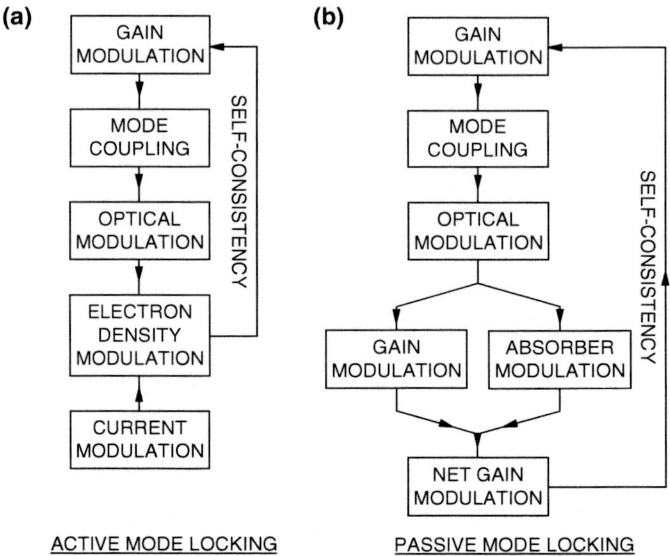

Fig. 9.1. (a) Self-consistent interpretation of active mode locking by current modulation of a semiconductor laser, (b) self-consistent interpretation of passive mode locking similar to that in (a). (From [88], ©1988 AIP. Reprinted with permission)

$$A_1 \left[\left(i\delta - \frac{1}{2\tau_\mathrm{p}} \right) (1 + b^2) + \frac{n_0 G\xi}{2} \right] = -\frac{nG\xi}{2} A_0, \qquad (9.3)$$

$$A_{-1} = A_1^*, \qquad (9.4)$$

where G is the optical gain constant, τ_p is the photon lifetime, b^2 is the gain difference between the central mode 0 and the neighboring ± 1 modes, $\delta = \Omega - \Delta\omega$, where $\Delta\omega$ is the frequency spacing between modes, and

$$\xi = \int u_0(z) u_{\pm 1}(z) w(z) \, \mathrm{d}z$$

is a geometric overlap factor, where $w(z)$ and $u_i(z)$'s are the spatial profiles of the modulated active medium and the optical modes, respectively. As is well known in analysis of conventional mode-locking analysis [94,95]. If the electron density modulation is distributed evenly in the cavity, orthogonality between different longitudinal modes leads to $\xi \to 0$ and no mode locking occurs. In fact, a general criterion for obtaining short optical pulses in mode locking is that the spatial distribution of the modulated portion of the active media be smaller than the physical extent of the optical pulse width, which explains the need for thin dye jets for femtosecond mode-locked dye lasers. In the situation under consideration here, where sinusoidal optical modulation at the round-trip frequency is the intended outcome, so that the active modulating section should *not* extend beyond approximately half of the cavity length. Under this circumstance the electron density can be reasonably approximated by a spatially averaged quantity. Substituting (9.3) and (9.4) into (9.2) results in

$$x^3 + x^2 b^2 + x \left[4(1 + b^2)^2 \delta^2 \tau_{\rm p}^2 - 2 \left(\frac{\xi n}{n_{\rm th}} \right)^2 \right] + 4(1 + b^2)^2 \delta^2 \tau_{\rm p}^2 b^2 = 0,$$

(9.5)

where

$$x = n_0/n_{\rm th} - (1 - b^2), \quad n_{\rm th} = 1/G\tau_{\rm p}.$$

(9.6)

The mode amplitudes are given by

$$A_1 = A_{-1}^* = \frac{-A_0 \xi n/n_{\rm th}}{2i\delta\tau_{\rm p} + x}.$$

(9.7)

The optical power output from the laser is proportional to the square of the field and is denoted as S:

$$S = A_0^2 + |A_1|^2 + |A_{-1}|^2 + A_0(A_1 + A_{-1}^*)e^{i\Omega t} + c.c.$$
$$\equiv S_0 + se^{i\Omega t} + c.c. \, .$$

(9.8)

9.1 Active Mode-Locking

The optical modulation interacts with the electron density via the rate equation

$$\dot{N} = \frac{J}{ed} - \frac{N}{\tau_{\rm s}} - GP_0(1 + p\cos\Omega t)N,$$

(9.9)

where $J = J_0 + j_1 \exp(i\Omega t)$ is the pump current density, $\tau_{\rm s}$, is the spontaneous lifetime, and $P_0 = \epsilon_0 S_0/2\hbar\omega$ is the photon density. In the limit of zero detuning ($\delta = 0$), the optical modulation depth p can be obtained from (9.5)–(9.8):

$$p = \frac{2|s|}{S_0} = 2\xi \frac{n}{n_{\rm th}} \frac{1}{b^2},$$

(9.10)

where a small modulation condition ($p \ll 1$) is assumed. The gain difference between the modes (b^2) is a function of the frequency separation between them and assuming a parabolic gain profile centered at mode number 0, $b^2 = (\Delta\omega/\omega_L)^2$, where ω_L is the width of the gain spectrum. A small-signal analysis of (9.9) with (9.10) gives the optical modulation response as a function of $\Delta\omega$:

$$p(\Delta\omega) = \frac{G\tau_{\rm p}j_1/ed}{GP_0 + (\Delta\omega^2/2\xi\omega_L^2)(i\Delta\omega + 1/\tau_{\rm s})}$$

(9.11)

The corner frequency of this function occurs at $(2\xi GP_0\omega_L^2)$. Using a typical value of $\omega_L = 2{,}500\,\mathrm{GHz}$ (which corresponds to a value of $b^2 = 2 \times 10^{-3}$ for a standard $300\,\mu\mathrm{m}$ cavity), assuming $\xi = 1/3$, and GP_0 is the inverse stimulated lifetime $= 1/(0.5\,\mathrm{ns})$, the corner frequency for $p(\Delta\omega)$ is $94\,\mathrm{GHz}$. The underlying reason for this very high frequency response can be found in (9.10). This relation shows that with a typically small value of b^2 encountered in semiconductor lasers, it is extremely easy for an electron density modulation to excite the side modes and hence results in an optical modulation (as long as the modulation frequency is nearly equal to the cavity mode spacing). The differential gain constant G is thus effectively amplified by a factor $1/b^2$, resulting in an extremely small equivalent stimulated lifetime which contributes to the high speed. The above results are based on a small-signal assumption, $p \ll 1$. It can be shown that as $p \to 1$ the available bandwidth will be substantially reduced so that short pulses generation is much harder than generating sinusoidal modulation at millimeter wave frequencies.

For finite detuning, one obtains from (9.5), (9.6) and (9.7)

$$A_1 = \frac{\xi(n/n_{\mathrm{th}})A_0}{(1 - 2i\delta\tau_{\mathrm{p}})(1 + b^2) - 1} = \frac{\xi(n/n_{\mathrm{th}})A_0}{b^2 - 2i\tau_{\mathrm{p}}\delta}. \tag{9.12}$$

The overall modulation response is shown in Fig. 9.2. The low-frequency portion is the usual direct modulation response of injection lasers. When modulated exactly at the cavity round-trip frequency $\Delta\omega$, the optical response depends on the value of $\Delta\omega$ as given by (9.11), represented by the dotted

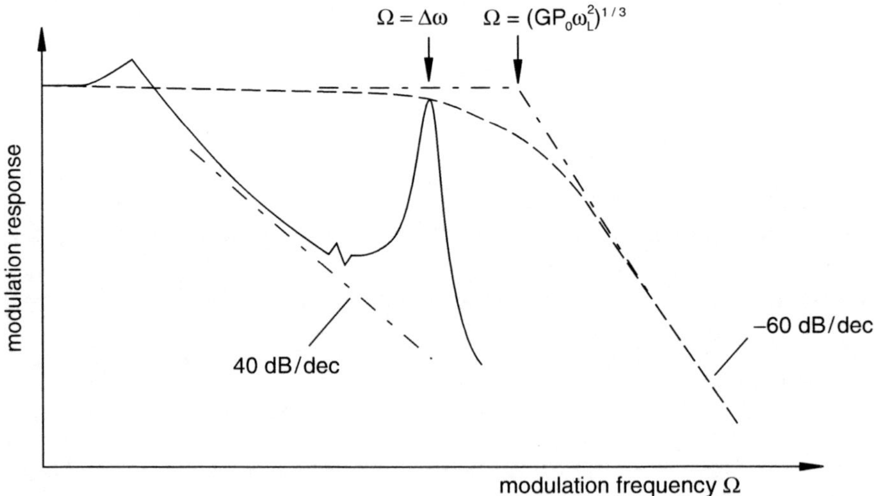

Fig. 9.2. Overall modulation response of an injection laser over the entire frequency range from baseband to beyond the inter-cavity-modal frequency. (From [88], ©1988 AIP. Reprinted with permission)

curve in Fig. 9.2. When the modulation frequency is detuned away from the cavity round-trip frequency, the response drops as a Lorentzian as given by (9.12).

The above analysis shows that there is basically no fundamental difficulty to produce sinusoidal optical modulation at the cavity round-trip frequency of up to 100 GHz if electrical parasitics were not a factor. This would be the case if the modulation is *internally* applied – i.e., passive mode locking using an intracavity saturable absorber – to be described in the next section. In this case, a weak, externally applied signal can serve to injection lock the self-optical modulation rather than creating the modulation itself.

9.2 Passive Mode-Locking

Assume that the intracavity absorber is formed by inhomogeneous pumping of a section of the laser diode. A self-consistent approach for passive mode locking, which is parallel to that of active mode locking is shown in Fig. 9.1b. In the presence of an absorber, one can define an equivalent electron density modulation n (similar to that used in the active mode-locking analysis above) such that $nG\xi$, equals the net gain modulation:

$$nG\xi \left(= \frac{n}{n_{\text{th}}\tau_{\text{p}}} \right) = n_{\text{g}}G_{\text{g}}f_{\text{g}} - n_{\text{a}}G_{\text{a}}f_{\text{a}}, \tag{9.13}$$

where n_{g} and n_{a} are the gain and absorber population modulation amplitudes governed by rate equations similar to (9.9), $G_{\text{g/a}}$ are the differential gain/absorption constants, and $f_{\text{g/a}}$ are geometric weighing factors. In the absence of externally applied modulation, the gain/absorption population modulation amplitudes are proportional to the optical modulation via

$$n_{\text{g}} = \frac{-G_{\text{g}}n_{\text{g0}}}{i\Omega + 1/\tau_{\text{g}} + G_{\text{g}}S_0}s,$$

$$n_{\text{a}} = \frac{-G_{\text{a}}n_{\text{a0}}}{i\Omega + 1/\tau_{\text{a}} + G_{\text{a}}S_0}s, \tag{9.14}$$

where the τ's are the spontaneous lifetimes and $n_{\text{g0/a0}}$ are the saturated steady-state electron densities in the gain/absorber regions. The optical modulation s is related to n via (9.5), (9.6), (9.7), and (9.8), so that (9.13) constitutes a self-consistent condition from which one can obtain δ and x, and subsequently the optical modulation amplitude. The result is

$$p = \sqrt{2\left[1 - \left(\frac{\Omega^2}{2\psi_{\text{r}}(\Omega)\omega_L^2}\right)^2\right]}, \tag{9.15}$$

where $\psi_{\text{r}}(\Omega)$ is the real part of the net gain modulation response [right-hand side of (9.13) normalized by s].

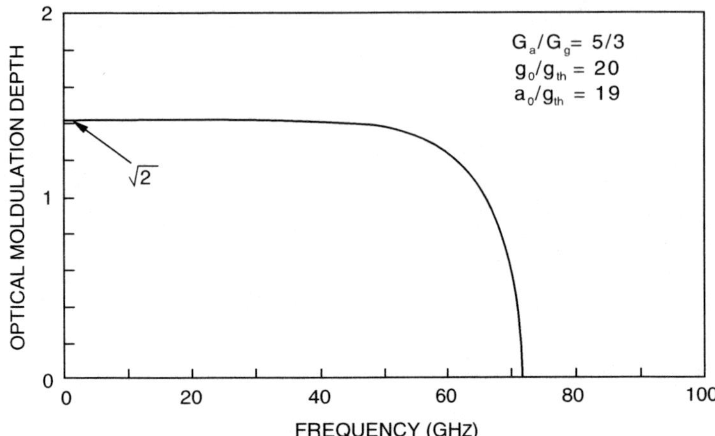

Fig. 9.3. Optical modulation depth of a *passively* mode-locked semiconductor laser as a function of cavity round-trip frequency (inverse of the cavity round-trip time). The parameters used are $g_0/g_{th} = 20$, $a_0/g_{th} = 19.5$, and $G_a/G_g = 5/3$. (From [88], ©1988 AIP. Reprinted with permission)

Figure 9.3 shows a plot of the optical modulation depth p as a function of passive mode-locking frequency Ω. At low frequencies, the optical modulation depth equals $\sqrt{2}$ and corresponds to the state when equal amount of power resides in the main mode as in the sum of both side bands. The apparent >100% modulation depth occurs because in (9.8) only the first harmonic is included, but for the purpose here it is sufficient to know that it is possible in theory to obtain optical modulation close to 100% up to very high frequencies until the cutoff point. A detailed analysis shows that the cutoff frequency depends on the amount of absorber and, most important, the ratio G_a/G_g. The maximum cutoff frequency for $G_a/G_g = 5/3$ is $f \approx 40\,\mathrm{GHz}$, whereas for $G_a/G_g = 5/1$ it can extend to $\gtrsim 160\,\mathrm{GHz}$, while for $G_a/G_g < 1$ mode locking is not possible, a well-known conclusion from the standard time domain theory [96–98]. The higher ratios can be realized in a saturable absorber with low saturation power, and can be attained with an inhomogeneously pumped single quantum well laser structures [99].

An experimental demonstration of ultra-high frequency passive mode-locking [100] is shown schematically in Fig. 9.4, in which a $250\,\mu\mathrm{m}$ long laser diode where the top contact was segmented into three parts, with the middle section reverse biased, thus acting as an absorber, while the two end sections are forward biased thus provide gain for the device. The fast optical output from this device was observed with an optical second harmonic generation (SHG) autocorrelation apparatus, which yields the time domain autocorrelation of the optical field output from the laser. If the optical field (intensity) output were of well-isolated periodic optical pulses, the SHG trace should look likewise. The width of optical pulses can be inferred from the width of

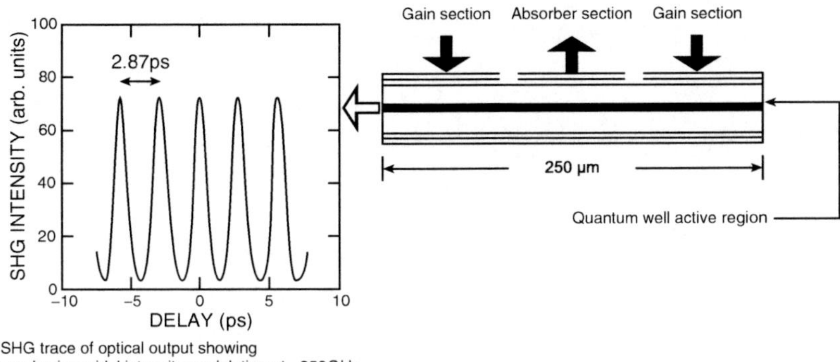

Fig. 9.4. Schematic diagram of a monolithic passively mode-locked laser (*right*), measured second harmonic autocorrelation trace of the optical output (*left*), showing almost sinusoidal intensity variation at ~350 GHz. (From [100], ©1990 AIP. Reprinted with permission)

the pulses in the SHG trace. On the other hand if the optical output from the laser assumes more of a sinusoidal variation in time, the SHG trace will resemble more of a sinusoidal shape, the period of the sinusoid-like SHG trace gives the frequency of the sinusoidal intensity output from the laser. In the case illustrated in Fig. 9.4 the frequency of the optical intensity oscillation was observed to be ~350 GHz, which corroborates well with the round-trip transit time of the 250 μm long laser cavity. This is a strong evidence of passive mode-locking as discussed in Sect. 9.2 above. However, the SHG trace as reported does not has the characteristic of distinct pulses, but rather that of a sinusoidal oscillation. This is evidence that the optical output from the laser is sinusoidally modulated rather than consists of distinct pulses, which is not expected at such a high mode-locking frequency of 350 GHz.

Performance of Resonant Modulation in the mm-Wave Frequency Range: Multi-Subcarrier Modulation

Optical transmitters capable of efficiently transporting several millimeter-wave (mm-wave) subcarriers to/from fiber-fed antenna sites in indoor/out-door mm-wave mobile/point-to-point wireless networks are of considerable importance in mm-wave free space links [101, 102]. Future deployment of a fiber infrastructure in these systems rests primarily upon the availability of *low-cost* mm-wave optical transmitters. Optical transmission of a *single* narrowband ($50\,\mathrm{Mb\,s^{-1}}$) channel at $45\,\mathrm{GHz}$ was demonstrated using resonant modulation of an inexpensive, conventional semiconductor laser with a baseband direct modulation bandwidth of $<5\,\mathrm{GHz}$ [103]. It was shown that this technique provides a means of building simple, low-cost, narrowband ($<1\,\mathrm{GHz}$) mm-wave subcarrier optical transmitters for frequencies approaching $100\,\mathrm{GHz}$. In this chapter, the *multichannel* analog and digital performance of these transmitters at a subcarrier frequency of $\sim 40\,\mathrm{GHz}$ are described. Two-tone dynamic range is characterized in detail as a function of bias to the laser, and a maximum dynamic range of $66\,\mathrm{dB\text{-}Hz^{-2/3}}$ is found. Although this is modest by conventional, say, CATV standard, it is adequate for serving a typical indoor picocell with a $40\,\mathrm{dB}$ variation in received RF power for a per-user voice channel bandwidth of $30\,\mathrm{kHz}$ and a carrier-to-interference ratio of $9\,\mathrm{dB}$. A multichannel system implementation of resonant modulation is also presented in which two signals centered around $41\,\mathrm{GHz}$ operating at $2.5\,\mathrm{Mb\,s^{-1}}$ BPSK are transmitted over $400\,\mathrm{m}$ of single mode optical fiber. The required RF drive power to the laser to achieve a bit-error-rate (BER) of 10^{-9} for *both* channels transmitting *simultaneously* is measured to be $<5\,\mathrm{dBm}$ per channel. Based on these transmission results and by taking advantage of conventional wireless time-division multiplexing techniques in which up to eight users can share a single channel [104], these mm-wave links are potentially adequate in remoting signals from an antenna serving up to 16 mobile users in an indoor environment.

The setup used to perform two-tone measurements and multichannel digital transmission test of the mm-wave optical transmitter is illustrated in Fig. 10.1. The laser used was a GaAs quantum-well laser with a cavity length

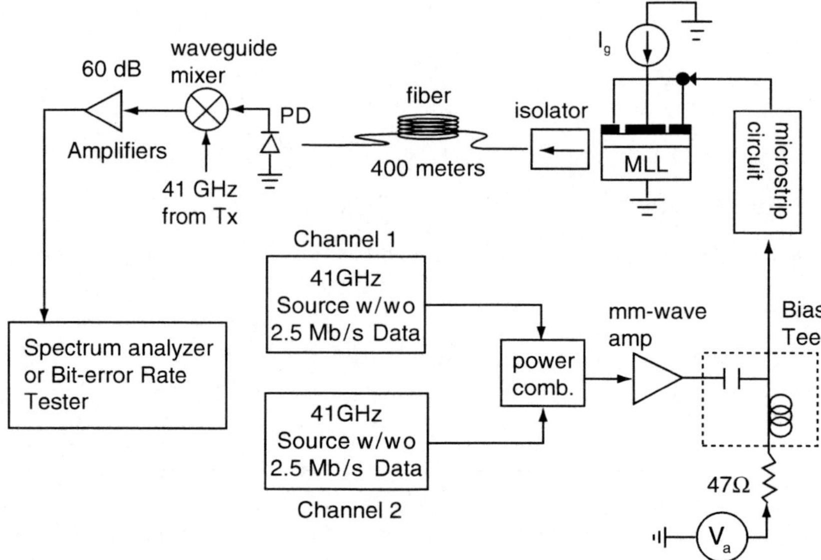

Fig. 10.1. Setup to measure the modulation response at 41 GHz, perform two-tone measurements, and characterize the digital performance of the transmitter (a multi-contact but otherwise conventional laser diode). (From [105], ©1995 IEEE. Reprinted with permission)

of $\sim$900 µm and emitting at 850 nm. First, the small-signal modulation response at the cavity round-trip frequency is measured. The modulation signal is delivered to the laser with the aid of a single-section microstrip matching circuit having a response shown in Fig. 10.2. The matching circuit reduces the reflection coefficient S_{11} of the laser to -15 dB at 41.15 GHz as measured. Modulation response around 41 GHz is shown in Fig. 10.3 for several bias conditions. By simply adjusting the bias to the laser, a higher modulation efficiency is achieved at the expense of passband bandwidth [103]. At a modulation efficiency of -5 dB (relative to that at dc) the passband bandwidth is $\sim$200 MHz.

For dynamic range measurements, two mm-wave tones from two Gunn oscillators operating at 41 GHz and separated by $\sim$1 MHz are electrically power-combined and delivered to the laser. Electrical isolation between the oscillators is $>$30 dB. The light emitted from the laser is sent through 400 m of single-mode fiber where it is detected, amplified, downconverted to IF and observed on a spectrum analyzer. The resulting dynamic range plots are shown in Fig. 10.4. A comparison of Figs. 10.3 and 10.4 reveals a trade-off between modulation efficiency and dynamic range. A maximum dynamic range of 66 dB-$\text{Hz}^{-2/3}$ is obtained for this laser. Note that the IP3 point is comparable to that below relaxation oscillation ($\sim$10 dbm). At a higher modulation efficiency, the dynamic range is reduced to $\sim$58 dB-$\text{Hz}^{-2/3}$ due to the increased, resonantly

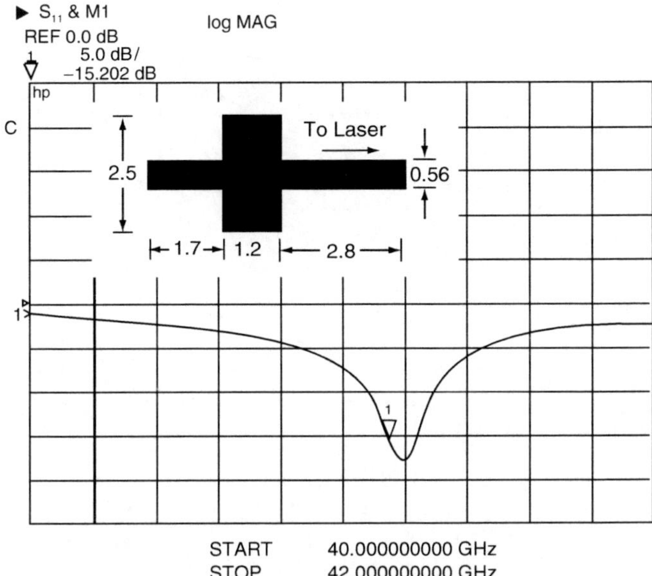

Fig. 10.2. Measured reflection coefficient S_{11} of the combined laser plus matching circuit. The mm-wave matching circuit was fabricated on a 0.18-mm-thick Duroid board with metallization dimensions (in millimeters) shown in the *inset*. (From [105], ©1995 IEEE. Reprinted with permission)

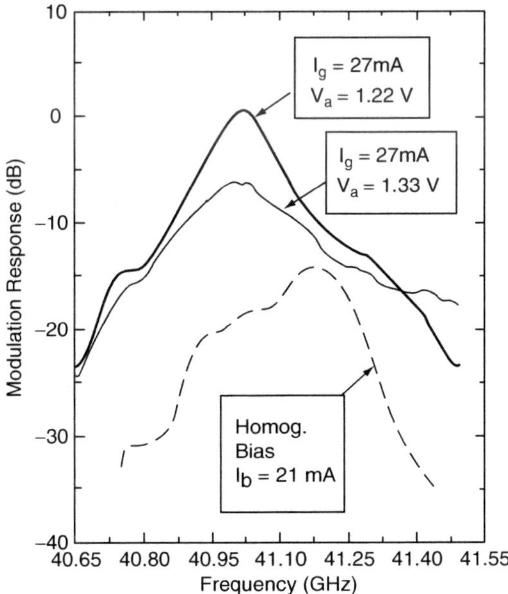

Fig. 10.3. Measured small-signal modulation response at the cavity round-trip resonant frequency of 41 GHz for various bias conditions. The vertical axis is relative to that of the dc of $0.26\,\mathrm{W\,A^{-1}}$. (From [105], ©1995 IEEE. Reprinted with permission)

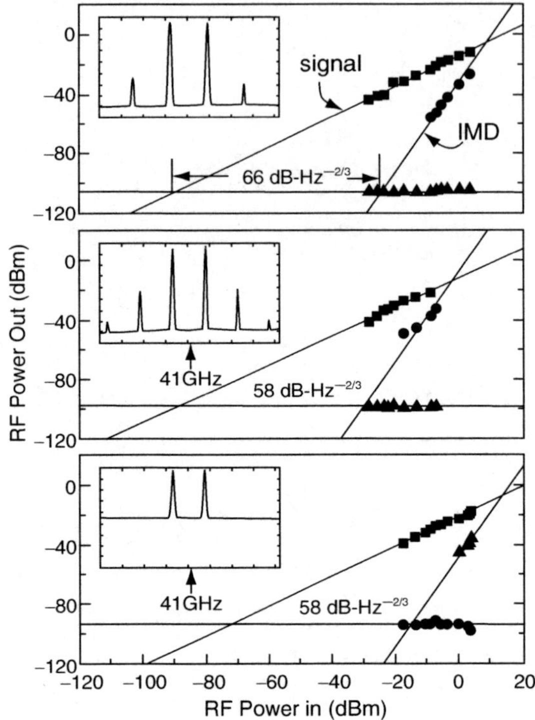

Fig. 10.4. Measured two-tone dynamic range under the same bias conditions used to obtain the modulation response measurements in Fig. 10.1. The *top figure* is for the bias condition $I_g = 27\,\text{mA}$, $V_a = 1.33\,\text{V}$; the *middle figure* $I_g = 27\,\text{mA}$, $V_a = 1.22\,\text{V}$; the *bottom* for homogeneous bias $I_b = 21\,\text{mA}$. The scale for the insets is $5\,\text{dB/div}$. RBW = 1 MHz. (From [105], ©1995 IEEE. Reprinted with permission)

enhanced noise. For homogeneous bias, the distortion level is lower (for the same drive power), but the corresponding high level of noise leads to a low dynamic range ($58\,\text{dB-Hz}^{-2/3}$). The insets show the measured intermodulation products for each bias condition at an electrical drive power per channel of $-6\,\text{dBm}$. No difference was observed when the dynamic range measurements were repeated for the same bias conditions in the absence of the 400 m of fiber, which suggests the dynamic range was limited by the laser. Improvement in the dynamic range can be achieved (at the expense of modulation efficiency) by incorporating intracavity frequency selective elements such as gratings or coupled cavities [106].

Next, the performance of the transmitter modulated by two binary-phase shift-keyed (BPSK) subcarrier channels is ascertained. The laser is biased for a modulation efficiency of $\sim$0 dB (relative to dc), a passband bandwidth of $\sim$200 MHz, and emitting an optical power of $\sim$2 mW. Two channels each transmitting pseudorandom ($2^9 - 1$) return-to-zero data at $2.5\,\text{Mb}\,\text{s}^{-1}$ are

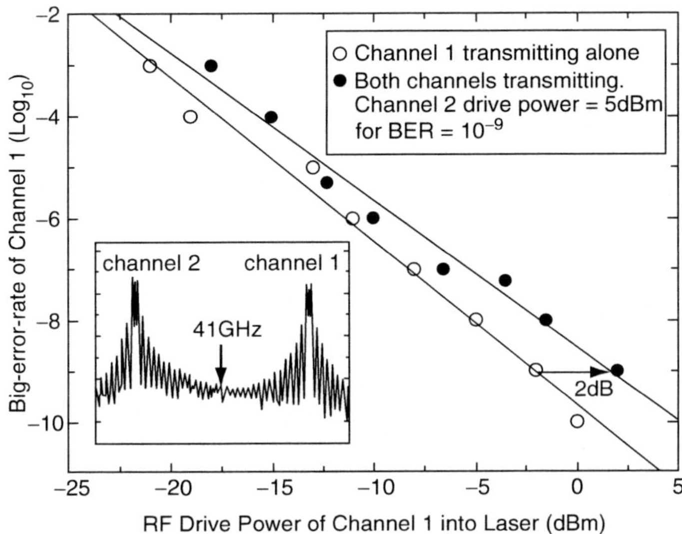

Fig. 10.5. Bit-error-rate of channel 1 as a function of RF drive power of channel 1 with and without the presence of channel 2 for $2.5\,\mathrm{Mb\,s^{-1}}$ retum-to-zero BPSK modulation centered around 41 GHz. The *inset* shows the received RF spectrum of both channels transmitting simultaneously after transmission over 600 m of single-mode optical fiber. The inset scale is 5 dB/div. (From [105], ©1995 IEEE. Reprinted with permission.)

upconverted to 41 GHz using a Q-band waveguide mixer and power combined to modulate the laser. The signals are transmitted over 400 m of single-mode fiber. At the receiver, the signals are down-converted to baseband, amplified and sent to an error-rate tester. The BER versus electrical drive power of channel 1 (centered at $\sim$41.15 GHz) is first measured with channel 2 (centered at $\sim$40.95 GHz) turned off as shown in Fig. 10.5. The RF power required to achieve a BER of 10^{-9} for this single channel is -2.5 dBm. With channel 2 activated, an additional 2 dB of RF power (power penalty) is required to keep channel 1 operating at 10^{-9}. Likewise, the drive power required for channel 2 to operate at 10^{-9} in the presence of channel 1 is 5 dBm. The difference in RF power between the channels for 10^{-9} operation stems from injection-locking effects which occur under higher RF drive power. Injection locking at channel 1 leads to a higher level of noise at channel 2, as illustrated in the inset of Fig. 10.4. At lower drive powers, both channels act independently as evidenced by the convergence of the BER curves at low drive powers.

The above sections have demonstrated the multichannel analog and digital performance of mm-wave optical transmitters based on resonant modulation of monolithic semiconductor lasers, and have established their feasibility as narrowband optical transmitters in fiber links serving remote antennae in in-door mm-wave wireless microcells. A two-tone dynamic range of 66 dB-Hz$^{-2/3}$

was obtained at a cavity round-trip frequency of 41 GHz and was limited by the high level of noise. Optical transmission over 400 m of fiber of two simultaneous $2.5\,\mathrm{Mb\,s^{-1}}$ channels centered 41 GHz and operating at $<5\,\mathrm{dBm}$ RF drive power per channel at a BER of 10^{-9} was also demonstrated.

Resonant Modulation of Single-Contact Lasers

Chapters 9 and 10 above describe in detail the concept and performances of narrow-band subcarrier modulation of a laser diode at mm-wave frequencies based on the concept of mode-locking, alternatively known as "resonantly-enhanced modulation" (or simply "resonant modulation" for short). This technique is useful in applications where optical fiber is used for the remote transport of millimeter wave ($>30\,$GHz) signals for phased-array antenna systems [107] and wireless personal communication networks [101]. Due to the fact that efficient longitudinal mode-coupling *requires inhomogeneous* modulation of the laser cavity in the *longitudinal spatial dimension*, previous demonstrations of monolithic resonant modulation such as those described in Chap. 10 were performed using split(multi)-contact laser diodes. Although the fabrication of a multi-contact structure itself is neither difficult nor intricate, the increased complexity is undesirable – considering the fact that it *does* require a *non-standard* fabrication process; rendering them unsuitable for seamless integration into a standard telecom laser fabrication line. Recent measurements of millimeter-wave propagation along the contact stripe of a semiconductor laser showed a high attenuation ($\sim 60\,$dB mm^{-1} at $40\,$GHz) of the signal along the laser stripe [108]. In this chapter, it is demonstrated that the confinement of high frequency modulation current resulting from the high signal attenuation along the length of the laser stripe can be utilized to achieve resonant modulation at $40\,$GHz of a *standard single contact* monolithic semiconductor laser. This concept is illustrated schematically in Fig. 11.1 for a ridge waveguide structure. The injected modulation current is confined to a local region near the feed point, resulting in a (longitudinally spatial) partial modulation of the laser cavity. The experimentally measured modulated light output and the small signal response of the laser device at two different feed points along the stripe of a semiconductor laser are shown in Fig. 11.2. In Fig. 11.3, a maximum modulation efficiency (on-resonance at $\sim 40\,$GHz) of $-20\,$dB (relative to that at baseband) can be observed. Also, using a simple distributed circuit model of the laser in conjunction with conventional

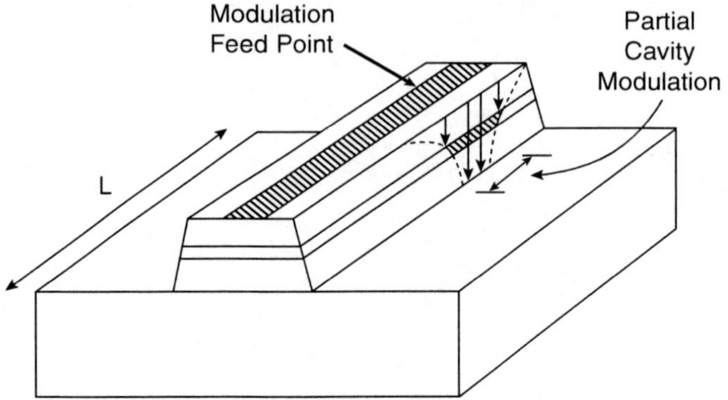

Fig. 11.1. Illustration of a single contact mode-locking experiment. A microwave probe is used to modulate the device at a particular feed point which leads to a partial modulation of the laser cavity. (From [109], ©1995 AIP. Reprinted with permission)

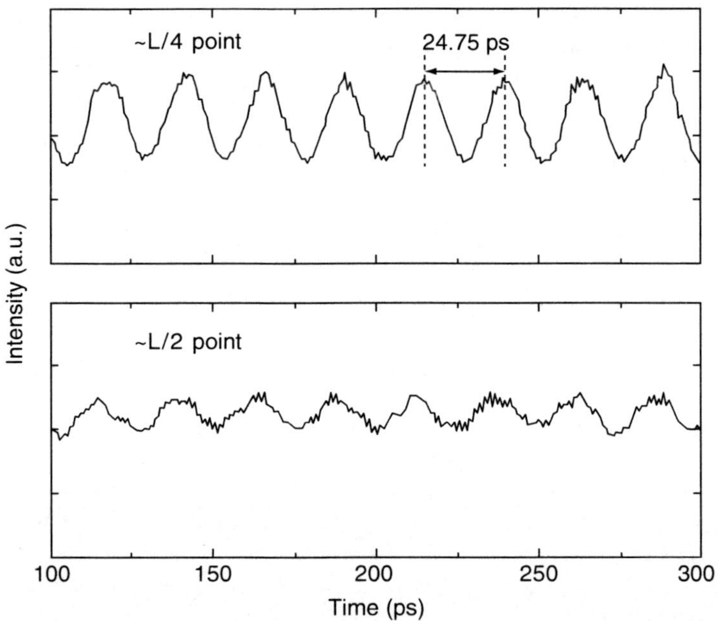

Fig. 11.2. Measured streak camera traces of the modulated light output from the single contact device at 40 GHz for two different microwave probe positions along the cavity. (From [109], ©1995 AIP. Reprinted with permission)

mode-locking theory, the characteristics and limitations of this technique are investigated.

The device used for the measurements is an InGaAs ridge waveguide laser with three quantum wells. The ridge structure is $4\,\mu$m wide, and the ground

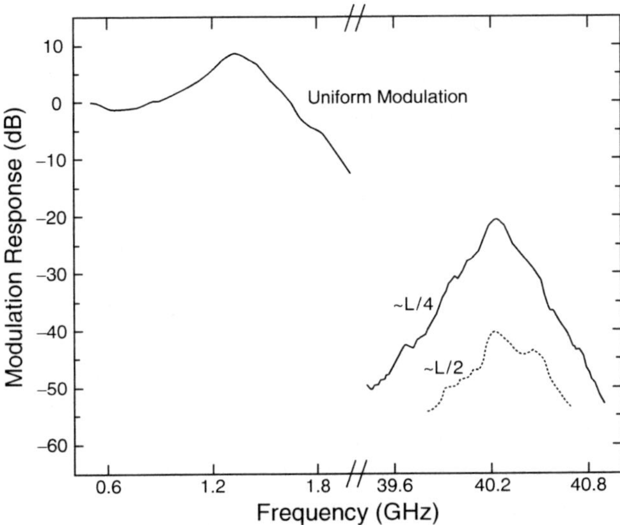

Fig. 11.3. The complete modulation response of the single contact device at low and high frequencies. The *solid curve* corresponds to an $L/4$ fed device, and the *dotted curve* corresponds to an $L/2$ fed device. The standard low frequency (baseband) relaxation resonance response can be observed when modulation current is applied uniformly over the length of the device. (From [109], ©1995 AIP. Reprinted with permission)

plane contacts are located $80\,\mu$m on both sides of the laser stripe. Although the device geometry is not perfectly coplanar, the $6\,\mu$m difference in height between the ridge and ground plane is small enough that a coplanar (ground–signal–ground) 40 GHz bandwidth probe can be used to inject signals into the device. For a cleaved length of $\sim$1,000 μm, the laser has a threshold current of 37 mA. In the first set of experiments, the optical output of the laser under millimeter-wave modulation was observed with a streak camera. The RF modulation was provided by a quadrupled 10.1 GHz synthesizer output, resulting in a 40.4 GHz signal driving the coplanar millimeter-wave probe. The streak camera was operated in synchroscan mode, and triggered with a phase locked 100 MHz output from the synthesizer. Figure 11.2 shows the optical modulation observed with a streak camera, measured with the probe positioned at $L/4$ and $L/2$ away from the edge of the laser. For both curves, the RF drive power and the bias current to the laser are the same. The streak trace clearly demonstrates millimeter-wave modulation of the light output at the "round-trip frequency" of 40.4 GHz. As expected, the mode-locking efficiency is substantially reduced when the laser is modulated near the center of the cavity ($\sim$L/2). Next, the small signal modulation response of the device at millimeter-wave frequencies was ascertained by sweeping the synthesizer over the frequency range of interest. At the receiver, the millimeter-wave modulated optical signal was detected with a high-speed photodiode followed by a down converting mixer driven at 39.0 GHz. The output of the mixer was amplified by 45 dB and observed on a RF spectrum analyzer. Figure 11.3 shows the small signal modulation response of the device at low frequencies and near the cavity round-trip frequency. For the quarter length ($\sim L/4$) fed device, the peak of the millimeter-wave response is 20 dB below the dc modulation efficiency, and the width of the passband is $\sim$160 MHz. These measurements are comparable to previous results obtained with split-contact lasers under homogeneous bias [103]. Again, note that the response of the center-fed device is small compared to the $\sim L/4$ case. These streak camera and RF response measurements substantiate the concept that efficient mode coupling is possible only under the condition of *non-uniform* current injection into the laser at the "round-trip frequency."

The modulation position dependence of the observed mode coupling is further investigated in a single contact device using the distributed circuit model used in [108] and shown in the inset of Fig. 11.4. The calculated amplitude of the normalized injection current into the laser is shown as a function of position along the device away from the feed point at 40 and 90 GHz. Notice that the amplitude of the injected current decreases rapidly with position, and drops to insignificant levels beyond 200 µm away from the feed point. For a round-trip frequency of 40 GHz, the laser is typically 1,000 µm long, resulting in a fractional cavity modulation of $\sim$20%. The modulation efficiency is addressed using a conventional mode-locking analysis along with the current distribution obtained from the distributed circuit model above. The

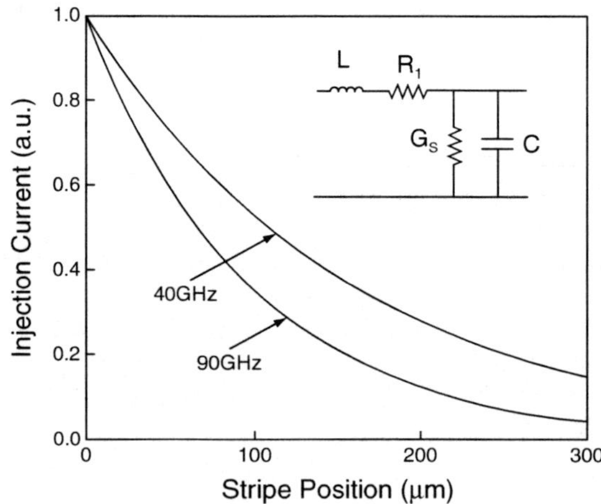

Fig. 11.4. Distribution of injected current into the active region as a function of length along the laser stripe at 40 and 90 GHz. The *inset* shows the distributed circuit model [108] used to calculate the current. (From [109], ©1995 AIP. Reprinted with permission)

self-consistent solution for active mode locking [110] shows that the optical modulation (p) is expressed as

$$p = 2\xi \left(\frac{n}{n_{\text{th}}} \right) \left(\frac{1}{b^2} \right),$$

(11.1)

where n is the amplitude of the time variation in the electron density, n_{th} is the threshold electron density, and the mode discrimination factor b is the ratio of the optical gain bandwidth to the round-trip cavity frequency. The parameter ξ is the overlap integral between adjacent longitudinal modes and the spatial variation of the gain modulation along the cavity. Since the modes are orthogonal, this amplitude is zero if the modulation is uniform over the cavity. If one assumes that the modulation in the photon density is small, such that the spatial dependence of the gain modulation can be approximated by that of the injection current, and that the cavity modes are given by solution to the one-dimensional Helmholtz equation, ξ can be expressed as

$$\xi = \frac{1}{L} \int_0^L I(z) \cos \frac{\pi z}{L} \, \mathrm{d}z,$$

(11.2)

where L is the laser length and $I(z)$ is the normalized spatial distribution of the modulation current. To determine $I(z)$, one can model the laser as a transmission line with two open ends. The impedance of an open ended transmission line is $Z = Z_0 \coth(\gamma l)$ where, Z_0 is the characteristic impedance of the stripline, γ is the complex propagation constant, and l is the distance from the probe to the open end [111]. Figure 11.5 shows this calculated mode-coupling amplitude as a function of position along the laser stripe for a "round-trip frequency" of 40 GHz. The maximum value of 0.06 is approximately a factor of 5 smaller than the maximum achievable value of ξ obtained when exactly one half of the cavity is uniformly modulated. For comparison, if one considers that the actual injection current profile being a delta function of amplitude $1/\gamma$ (i.e., no spreading of the modulation current), then (11.2) reduces to $\xi = (1/\gamma L) \cos \pi z/L$ which is shown as the dotted curve in Fig. 11.5. The similarity of this curve to the exact numerical calculation highlights the fact that the extent of the modulation current is very localized, and that ξ is proportional to $(\gamma L)^{-1}$. Hence, the value of ξ can be optimized by designing the transmission line for a particular value of γ. Note that the optimum probe position is approximately at $L/8$ away from the laser facet. In fact, the conclusion that $\sim L/8$ is the optimum feed point remains true over the entire millimeter-wave range (30–100 GHz). The maximum value of ξ also remains approximately constant over this range. This is understood by realizing that although the signal propagation is reduced at higher frequencies, the resonant device length is shorter, resulting in a comparable fractional cavity modulation. These results show that modulation of single contact monolithic semiconductor lasers is possible over the entire millimeter-wave frequency range up to 100 GHz.

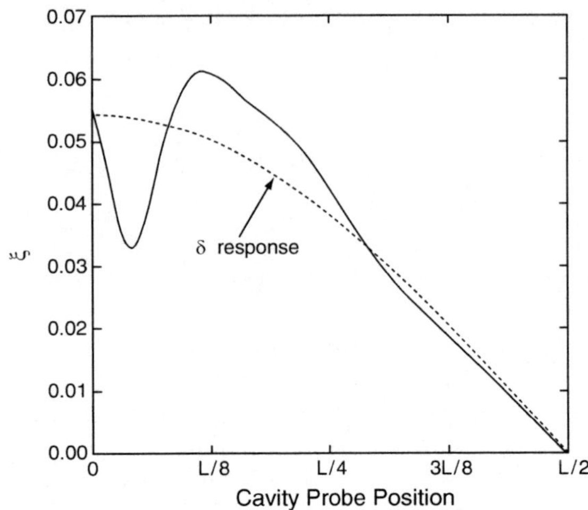

Fig. 11.5. Calculated mode coupling factor as a function of probe position for a device with a round-trip cavity frequency of 40 GHz. The *dotted curve* corresponds to the limiting case when the injected current is a delta function. (From [109], ©1995 AIP. Reprinted with permission)

The above results demonstrate the efficient modulation of a single contact monolithic semiconductor laser at the "round-trip frequency" of 40 GHz. Both temporal and RF measurements show clear evidence of millimeter-wave modulated light output from the laser. The observed feed point dependence of the modulation efficiency provides definitive evidence that mode-locking exists in the single contact device due to the limited propagation distance of the millimeter-wave signal. This technique was studied using a simple distributed circuit model of the laser in conjunction with conventional mode-locking theory. The optimum feed-point and mode-coupling factor over the millimeter-wave frequency range are found to be $\sim L/8$ and $(\gamma L)^{-1}$, respectively. These results are a very important step toward the realization of practical millimeter-wave optical transmitters based on direct modulation of monolithic semiconductor lasers.

Part III

Fiber Transmission Effects, System
Perspectives and Innovative Approach
to Broadband mm-wave Subcarrier
Optical Signals

12

Fiber Chromatic Dispersion Effects of Broadband mm-Wave Subcarrier Optical Signals and Its Elimination

12.1 Effects on Multichannel Digital Millimeter-Wave Transmission

The millimeter-wave (mm-wave) frequency band offers the *free-space* bandwidth necessary for future broadband wireless communications services. A high-capacity broadband wireless network can be the quickest and most cost-effective method of delivering services to a large number of customers in a dense environment. Millimeter-wave optical fiber links can effectively distribute mm-wave signals from a central office to remote antennas located at suitable vantage points for line-of-sight interconnection to other nodes of the network. As described in [112], these fiber links offer simplification of base stations and centralized control and stabilization of mm-wave carrier signals for conformity to FCC standards. Even though it is expected that this type of fiber systems will take advantage of legacy metropolitan fiber cable plant infrastructure at the dispersion minimum of 1,300 nm. The low fiber loss and availability of optical amplifiers at 1,550 nm can extend the central office coverage over a much larger service area than 1,300 nm links. Therefore it is still important to understand how dispersion in a fiber link can affect the transmitted information on mm-wave subcarriers. The effects of fiber chromatic dispersion on a single carrier have been examined in [113, 114]. A two-tone analysis was done in [115]. Because future broadband high-capacity services will have many digital channels, a multiple-channel analysis is needed. A CATV band simulation was reported in [116]. This chapter explores the effects of fiber chromatic dispersion on broadband 18 channel mm-wave subcarrier multiplexed (SCM) transmission. Instead of studying a particular digital QAM format, the study here concentrates on the fundamental limits due to chromatic dispersion-induced carrier degradation and intermodulation distortion at mm-wave frequencies. Transmission of multichannel mm-wave signals over single-mode fiber will also be limited by the optical link noise contributions from the receiver, laser RIN, and fiber amplifiers.

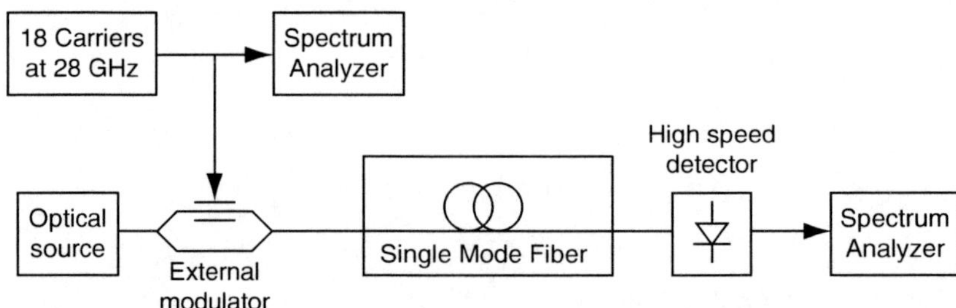

Fig. 12.1. 1,550 nm m-wave fiber system model. (From [117], ©1996 IEEE. Reprinted with permission)

A 1,550 nm *externally modulated* fiber system will be studied because

1. High frequency external optical modulators are available both commercially and in the laboratory (for more discussions, see Appendix C);
2. Chirping is minimal for these optical transmitter sources, the transmitter output is thus a pure amplitude modulation with very little phase modulation.

The system was modeled using the Signal Processing Worksystem simulation tool [118]. The model block diagram is shown in Fig. 12.1. A high-speed external optical modulator modulates a narrow linewidth (DFB) 1,550 nm optical source. The finite laser linewidth can lead to a negligible mm-wave carrier power degradation for typical line widths [119]. The modulator is modeled as linearized over the mm-wave band of interest (e.g., 27.5–28.5 GHz). External modulators have been demonstrated in this frequency range [120]. There are various methods for modulator linearization [121]. The electrical input to the optical modulator is the sum of 18 millimeter-wave channels.

The optical fiber is modeled as a unity amplitude, linear group delay filter. Standard single-mode fiber is used with a dispersion parameter of $18 \, \text{ps km}^{-1} \, \text{nm}^{-1}$ at 1,550 nm. Nonlinear fiber effects described in [122] are neglected in this simulation. A magnitude squared function models a high-speed detector and a FFT function gives the detector output spectrum.

A detected single subcarrier will experience a signal power variation with fiber transmission distance due to chromatic dispersion [113, 114]. This is because the single subcarrier is transmitted through the fiber as optical sidebands on the optical carrier. The sidebands experience phase changes due to fiber chromatic dispersion so that the detected signal is effectively a sum of two signals with a phase difference that is a function of the fiber length. It can be shown that for small modulation depths the detected signal power of the single subcarrier is approximately proportional to [114]

$$P = \cos^2 \left(\frac{\pi D \lambda^2 L f^2}{c} \right) \qquad (12.1)$$

where D is the dispersion parameter, L is the fiber length, and f is the frequency of the modulating signal. For $f = 28\,\text{GHz}$, $\lambda = 1,550\,\text{nm}$, and $D = 18\,\text{ps}\,\text{km}^{-1}\,\text{nm}^{-1}$, the maxima occur at multiples of $L = 8.85\,\text{km}$.

The first null does not represent the fundamental limit because the maxima are periodic at multiples of 8.85 km in this case. As long as the fiber length is adjusted properly, single subcarrier transmission is not limited by chromatic dispersion.

Chromatic dispersion will cause intermodulation distortion between multiple channels as the fiber distance is increased. A two-tone analysis was done in [115]. The analysis was a worst case analysis because the carriers were assumed to be in phase. Eighteen carriers are transmitted and the initial carrier phases can be set to zero or randomly generated.

The optical fiber affects the multichannel spectrum in two ways. First, there is the signal level change versus fiber length and frequency for each channel described above. Each frequency has a different periodic length. Second, the chromatic dispersion causes intermodulation distortion between the channels.

The center of the transmission band was chosen to be at 28 GHz which corresponds to the local multipoint distribution service (LMDS) band. To reduce aliasing in the spectrum, the sampling rate was set to 1,792 GHz and the frequency resolution was set to 27.34 MHz. A total of 19 channels at a channel spacing of 54.68 MHz (twice the resolution) gives a 980-MHz wide spectrum centered at 28.0 GHz, which was chosen to model a broadband and high-capacity communications system.

Channel 10, the center channel at 28.0 GHz where the distortion is highest, was removed and the remaining 18 tones were summed and transmitted to measure the intermodulation distortion. The optical modulation index (OMI) per channel was chosen to be 5.5% to prevent the total possible modulation from exceeding 100%.

As a reference Fig. 12.2a shows the simulated input spectrum with no dispersion. Figure 12.2b–d shows the simulated output spectra for the case of the 18 carriers initially in phase for the fiber lengths around 53 km. The fiber length $L = 53.09\,\text{km}$ for Fig. 12.2c corresponds to the sixth maxima of (12.1) at 28.0 GHz. The power is normalized to the fiber loss to isolate the dispersion effects. Any additional signal loss due to (12.1) can be compensated for by using erbium-doped fiber amplifiers.

The carrier-to-interference ratio (CIR) was defined as the ratio of the adjacent channel power to the intermodulation distortion power at 28.0 GHz. The adjacent channel power was the lower of Channel 9 (27.9453 GHz) and Channel 11 (28.05468 GHz). For the 49.8–56.2 km fiber length range, the CIR had a peak value of 41.5 dB and a minimum of 25.5 dB for the case of in-phase carriers. This CIR is well above the requirements for digital modulation formats such as QPSK and 16-QAM.

A more realistic case is when the carrier phases are uncorrelated. The interference level and adjacent channel power of 10 simulation runs with random

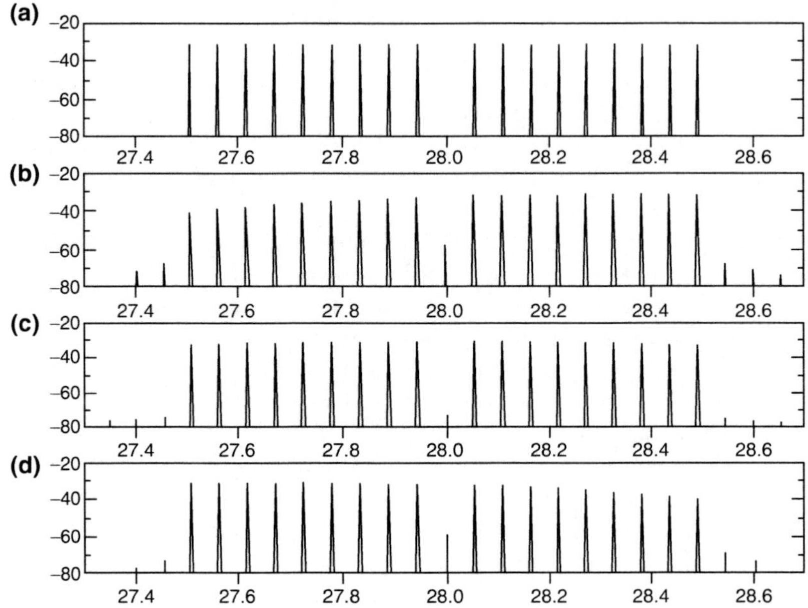

Fig. 12.2. Calculated output spectra after different fiber lengths of fiber with carriers in phase: (**a**) 0 km, (**b**) 51.4 km, CIR = 25.5 dB, (**c**) 53.1 km, CIR = 41.5 dB, and (**d**) 54.6 km, CIR = 26.7 dB. [*Vertical axis*: Normalized power (dB), *Horizontal axis*: Frequency (GHz)]. (From [117], ©1996 IEEE. Reprinted with permission)

initial phases for the 18 tones were averaged. The CIR was higher than the in-phase case by over 22 dB.

Figure 12.3 shows the simulated CIR for the in-phase carriers case versus fiber length at three different fiber-length ranges. As stated above, the case of random carrier phases gives an average improvement in CIR of >22 dB at each fiber length over the entire fiber length range. Therefore, Fig. 12.3 represents a worst-case lower limit on the CIR. The average CIR at each fiber length in the range is expected to have an average value of at least 42 dB with random carrier phases.

Figure 12.4 shows the maximum and minimum channel power for the three fiber-length ranges. The signal power variation due to fiber dispersion indicates how far the received channels are from the ideal situation of equal amplitude carriers at the receiver. The operational range is chosen to be fiber lengths which had a maximum channel power variation of 10 dB. With this requirement, there are 11 operational ranges for the fiber lengths. Figure 12.5 gives a graphical representation of the acceptable fiber lengths. Note that the fiber length ranges are approximately centered at multiples of 8.85 km.

Assuming a fiber loss of 0.25 dB/km at 1,550 nm and an erbium-doped fiber amplifier (EDFA) gain of 10 dB, two EDFA's are needed for 80-km transmission at 28 GHz and a received power of 0 dBm. Assuming a noise figure of

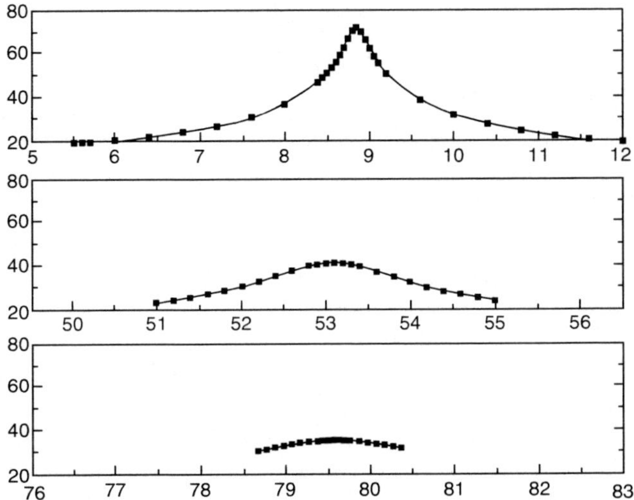

Fig. 12.3. CIR (dB) versus fiber length (km) for the case of carriers in phase. Random carrier phases give an average improvement of over 22 dB at each fiber length. (From [117], ©1996 IEEE. Reprinted with permission)

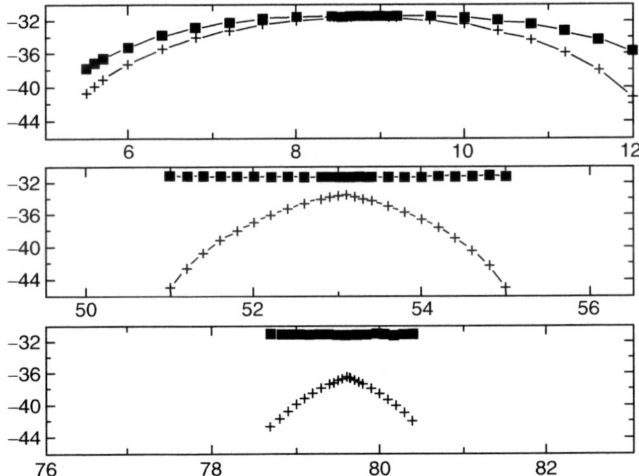

Fig. 12.4. Maximum channel power (*top curve*) and minimum channel power (*bottom curve*) in decibels versuo fiber length (km). (From [117], ©1996 IEEE. Reprinted with permission)

10 dB per amplifier, the amplifier noise limits the received CNR per channel to over 37 dB in the signal-spontaneous beat noise limit (additive noise from each EDFA) [123]. Consider a distribution tree system with two cascaded stages where each stage consists of an EDFA, a 10-way splitter, a second EDFA, and 40 km of fiber. One hundred base stations/fiber nodes at a distance of 80 km

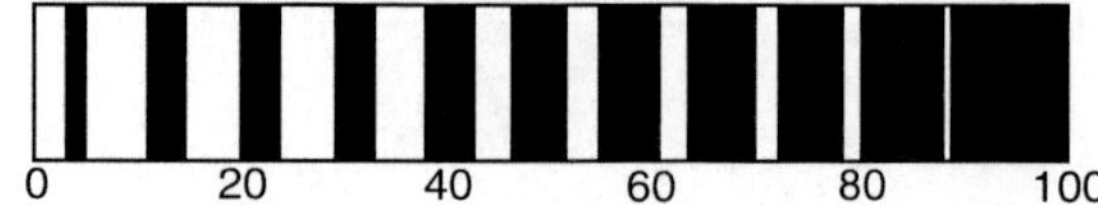

Fig. 12.5. Acceptable fiber length (*unshaded*) in km. (The *shaded regions* have a worst case CIR < 20 dB and/or an amplitude variation of over 10 dB). (From [117], ©1996 IEEE. Reprinted with permission)

away from the transmitter can be served with this system. With a 10-dB gain, a 10-dB noise figure, and a +10-dBm output power for each EDFA, the CNR per channel at the receiver in this case is limited to around 34 dB. This CNR is still sufficient for high fidelity transmission of QPSK and even 16-QAM [124].

Chromatic dispersion effects of standard single-mode fiber on the transmission of multichannel signals at millimeter-wave frequencies have been simulated using multiple unmodulated carriers similar to those used in typical distortion experiments [125]. Note that at 1,300 nm or at CATV frequencies these effects are small for practical fiber distances. Optical amplification can easily make up for any additional signal loss due to (12.1).

In conclusion, these computer simulation results can have a significant implications on the system architecture. Now a single central office or headend location can use up to an 88-km 28-GHz distribution tree or ring with optical amplifiers to distribute broadband QPSK and QAM information at millimeter-waves to an extensive service area. Fiber chromatic dispersion divides the fiber length into operation ranges. These ranges depend on the frequencies and system bandwidth, as well as the wavelength and dispersion parameter.

12.2 Elimination of Fiber Chromatic Dispersion Penalty on 1,550 nm Millimeter-Wave Optical Transmission

As described in the last section, millimeter-wave optical fiber distribution links between a central office and remote antenna sites provide centralized control and stabilization of the mm-wave signals and simplification of the remote electronics [112].

Conventional high-speed linear intensity modulators allow transmission of broadband, multichannel mm-wave signals [120]. Optical amplifiers at 1,550 nm can compensate for fiber attenuation and splitting losses to significantly extend the distribution range and coverage area of a single transmitter. However, recent work has shown that fiber chromatic dispersion, especially at 1,550nm, can cause severe signal power penalties at certain fiber distances and modulation frequencies [114, 117, 126, 127].

For conventional intensity modulation of a single-mode laser, symmetrical sidebands are created on the optical carrier. Owing to fiber chromatic dispersion, these sidebands experience relative phase shifts which depend on the wavelength, fiber distance and modulation frequency. Each sideband mixes with the optical carrier in the optical receiver. If the relative phase between these two components is $sim 180°$, the components destructively interfere and the mm-wave electrical signal fades. The detected signal power variation is proportional to (12.1) where D is the dispersion parameter, L is the fiber length, and f is the modulation frequency [114, 117, 126].

However, by simply filtering out one of the optical sidebands, the problem is eliminated. Now dispersion only adds a phase shift with no amplitude change. At mm-wave frequencies of 25–60 GHz, the optical sidebands are separated by 0.2–0.5 nm from the optical carrier at 1,550nm. A *fiber Bragg grating* provides a simple and narrowband commercially available notch filter that can be tailored to the laser wavelength. An external optical modulator with the fiber Bragg grating sideband filtering effectively produces single-sideband optical modulation at mm-wave frequencies. Note that the filter requirements are relaxed as the modulation frequency increases. Single-sideband filtering using a Mach–Zehnder-type optical filter was used to reduce dispersion effects on a high-speed baseband digital modulation signal in [128].

To demonstrate the dispersion penalty and its reduction, an SDL (now a division of JDS Uniphase) 1,556.5 nm DFB laser with additional optical isolation was modulated with a 50 GHz Mach–Zehnder LiNbO$_3$ intensity modulator [120], as shown in Fig. 12.6. The modulator was biased at the half-intensity point resulting in 1 mW of optical power at the modulator output. The modulator electrical drive signal was a swept frequency tone from an HP 40 GHz synthesized sweeper. The optical modulator output was sent through 51.1km of standard non-dispersion shifted single-mode fiber (Corning SMF-28) and detected using a Bookham, Inc. 40 GHz high-speed photodetector.

Figure 12.7 shows the optical spectrum with and without the optical filter with a modulator electrical drive frequency of 40 GHz. The optical modulation depth was ∼8%. The 40 GHz modulation produces upper and lower optical sidebands spaced 0.32 nm from the optical carrier. The filter attenuates the upper optical sideband by ∼22 dB.

Figure 12.8 shows the received signal power versus frequency with and without the optical filter as the modulator drive is varied from 30 to 40 GHz. The power is normalized to the received signal power at 0 km without the optical filter and fiber loss. Without the filter, there are signal nulls at certain

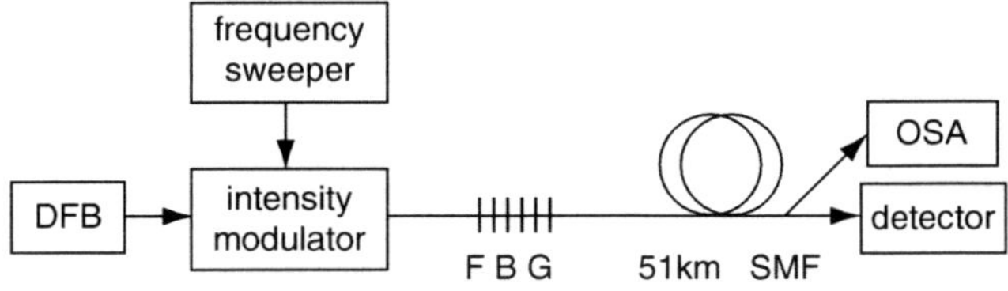

Fig. 12.6. Experimental millimeter-wave setup. (From [129], ©1997 IEE. Reprinted with permission)

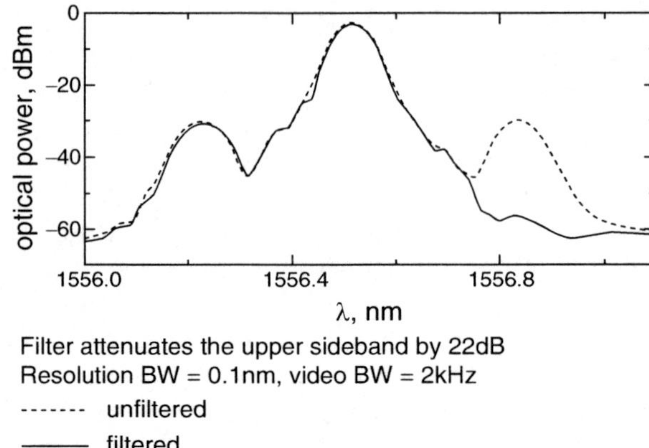

Fig. 12.7. Unfiltered and filtered optical spectra with 40 GHz modulation. (From [129], ©1997 IEE. Reprinted with permission)

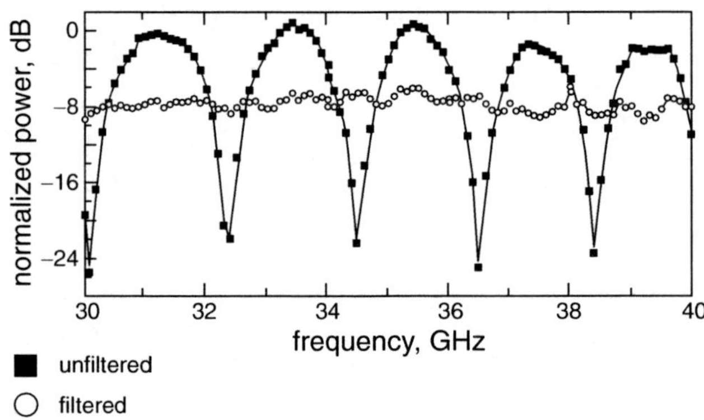

Fig. 12.8. Unfiltered and filtered frequency response at 51.1 km. (From [129], ©1997 IEE. Reprinted with permission)

frequencies due to fiber dispersion, according to (12.1). However, with the optical filter, the nulls are eliminated over the entire frequency range. Note that when one sideband is filtered, half of the optical sideband power is removed which results in a 6 dB electrical loss relative to the maximum electrical signal power when no filter is used, as shown in the measurements of Fig. 12.8. An additional 2 dB loss due to the filter insertion loss results in the 8 dB relative electrical loss shown in Fig. 12.8. In distribution links, this loss can easily be compensated for using EDFAs.

The signal power was also measured as the fiber length was varied. Standard non-dispersion shifted single-mode fiber (Corning SMF-28) was available in roughly 2.5 km increment reels up to a total fiber length of 51.1 km. The modulation frequency was lowered to 25 GHz so that the power nulls given by (12.1) were separated by a large enough fiber distance to be resolved by the

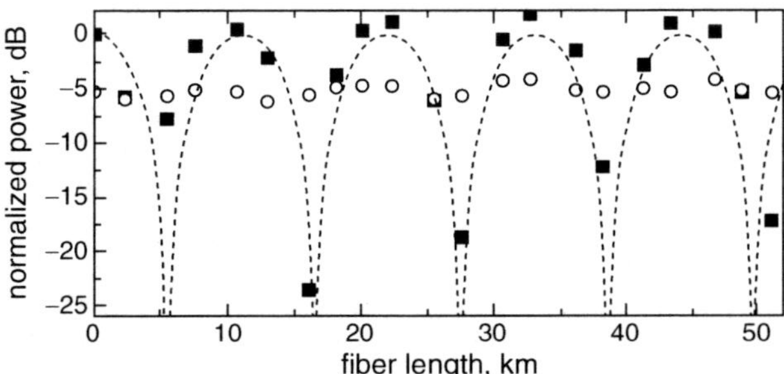

Fig. 12.9. Unfiltered, filtered and theoretical response against fiber length with 25 GHz modulation. (From [129], ©1997 IEE. Reprinted with permission)

2.5 km fiber length resolution. The measurements are shown in Fig. 12.9. The theoretical curve according to (12.1) is also shown. A value of $18\,\mathrm{ps\,km^{-1}\,nm^{-1}}$ was used for the dispersion parameter D in (12.1). Without the filter, there are five nulls over the 51 km fiber length range. With the filter, there are no frequency nulls over the entire fiber range.

This chapter demonstrates a simple method for the elimination of the fiber dispersion penalty on conventional external optical intensity modulation at millimeter-waves using a fiber Bragg grating filter to produce single-sideband optical modulation. It is shown that this simple filtering technique removes the signal power variation over modulation frequency and fiber length while introducing a fixed signal loss due to the removal of one optical sideband. The single-sideband optical modulation still allows broadband, multichannel millimeter-wave transmission. Although only 51 km of fiber was available, this filtering technique is not limited to this distance and much longer transmission distances should be possible. It is also interesting to note that the filter can be used at the optical receiver instead of at the transmitter with the same results.

13

Transmission Demonstrations

13.1 1,550 nm Transmission of Digitally Modulated 28-GHz Subcarriers over 77 km of Non-Dispersion Shifted Fiber

In the highly competitive race to provide two-way broad-band network access to the home/enterprise, minimizing the time-to-market is essential. Millimeter-wave (mm-wave) wireless systems, unlike wired systems, can be quickly set up on-demand, to provide two-way connections between premises and base station nodes. Millimeter-wave optical fiber distribution links between a central office and remote base stations provide centralized control and stabilization of the mm-wave signals and simplification of the remote electronics [112].

As explained in the Preface of this book this kind of distribution system for implementation of a mm-wave free-space network is most cost-effective, and expedient to bring on-line, if legacy infrastructures of fiber network installed for telecom/metropolitan networks can be utilized through leasing of dark fibers. Metropolitan networks consist mostly of single mode fibers at the dispersion minimum of 1.3 µm, while telecom networks consist mostly of single mode fiber at the loss-minimum wavelength of 1.55 µm. The low fiber loss and availability of optical amplifiers at 1.55 µm allow a central office location to extend its coverage over a much larger service area than using 1,300 nm metropolitan fibers alone. However, fiber chromatic dispersion has a significant effect at 1.55 µm and has been studied in [113–117, 130]. In [117], multichannel transmission in the local multipoint distribution service (LMDS) band at 28 GHz using an external optical modulator was simulated. Despite the dispersion-induced mm-wave signal degradations, the simulation results suggest that acceptable carrier-to-interference ratios (CIR) are possible up to 80 km for unmodulated carriers. Transmission of a single mm-wave channel over long spans of fiber at 1,550 nm has been demonstrated using heterodyne techniques and optical amplification [127, 131]. This chapter describes

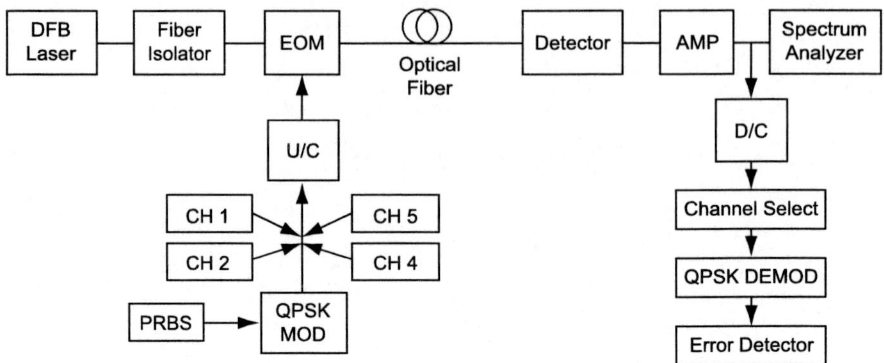

Fig. 13.1. Experimental setup for signal and distortion measurements. (From [126], ©1997 IEEE. Reprinted with permission)

experimentally observed fiber chromatic dispersion effects on multichannel digital millimeter-wave transmission, as well as transmission of multiple mm-wave carriers including QPSK digital modulation at 28 GHz over long lengths of fiber without optical amplification.

Other than direction modulation of (resonantly-enhanced) laser diodes, an alternative method of achieving optical modulation at mm-wave frequencies is by modulating the output of a (continuously operated) optical source with a high-speed optical modulator as in [117]. This chapter describes computer simulation results that fiber chromatic dispersion causes two effects on multichannel mm-wave transmission: (1) an amplitude variation versus frequency and distance and (2) intermodulation distortion between channels. The experimental setup for multichannel mm-wave transmission is shown in Fig. 13.1. The broad-band mm-wave transmitter consists of commercially available components – a packaged Ortel DFB laser with a wavelength of 1,543 nm with additional fiber isolation was modulated using a 30 GHz Sumitomo Cement LiNbO$_3$ external optical intensity modulator (EOM) with a half-wave voltage of $V_\pi = 4.7$ V. The EOM was biased at the half-intensity point and was not linearized. The optical power out of the EOM was about 0.7 mW.

A total of five mm-wave channels could be used to drive the EOM. The channel spacing was chosen to be 20 MHz. Four of the channels were unmodulated carriers from frequency synthesizers at 1,960, 1,980, 2,020, and 2,040 MHz which were power combined, upconverted, amplified and filtered to provide four mm-wave carriers at 27.96, 27.98, 28.02, and 28.04 GHz. The upconverter consisted of a 44-GHz Watkins–Johnson mixer with a 26-GHz LO from an HP83650A 50-GHz Synthesized Sweeper, and two HP mm-wave amplifiers that provide 38 dB of gain to drive the EOM. The fifth channel was a 5 Mbaud QPSK digitally modulated and raised-cosine filtered carrier at 2,000 MHz that was upconverted to 28.00 GHz, the center channel. An optical modulation depth (OMD) of 24% *per* channel was chosen to give an rms OMD of 38%. This OMD was chosen because the EOM was not linearized and thus the need to limit clipping distortion [132].

Standard non-dispersion shifted single mode fiber (NDS-SMF) with minimum attenuation at 1,550 nm and minimum disperion at 1,310 nm was used and varying lengths up to 76.7 km were available. At 1,543 nm the fiber loss is $0.2\,\mathrm{dB\,km^{-1}}$ and the dispersion parameter is about $17\,\mathrm{ps\,km^{-1}\,nm^{-1}}$.

At the receiver end, the detector was a 40-GHz Bookham, Inc. high-speed photodiode with a responsivity of 0.1 A/W at 28 GHz. A Hewlett-Packard low noise mm-wave amplifiers provided 35 dB of gain at 28 GHz with a 4-dB noise figure. Signal and intermodulation powers were measured on an HP8565E Spectrum Analyzer. For bit-error-rate (BER) measurements, the mm-wave spectrum was downconverted using a 44-GHz Watkins–Johnson mixer with a 28.275-GHz LO from an HP83650A 50 GHz Synthesized Sweeper, and the QPSK channel was filtered and demodulated. The demodulator had AGC, equalization and error correction. The recovered bit stream was input to an HP3764A Error Detector.

Using a single mm-wave source to drive the EOM, the frequency was swept from 27.5 to 28.5 GHz. The detected signal power normalized to the received power at 0 km and to the fiber loss is shown in Fig. 13.2 at three different lengths of fiber around 75 km. In the absence of fiber chromatic dispersion there should be no power penalty and all of the curves would be straight lines at 0 dB. However, dispersion causes the signal power variation versus frequency shown in Fig. 13.2. The intensity modulation of the single-mode laser with a single mm-wave subcarrier creates upper and lower sidebands on

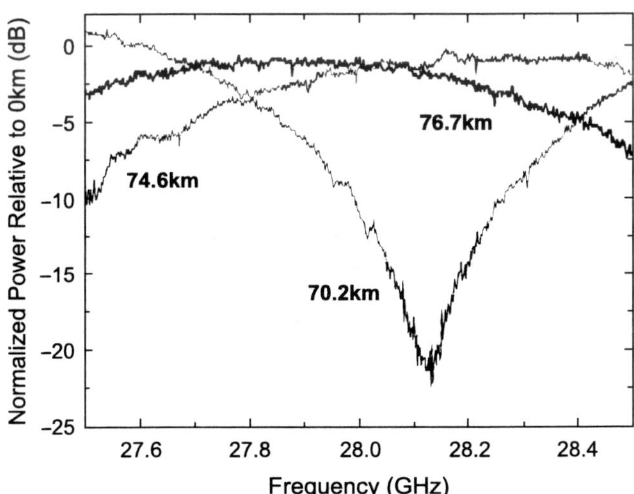

Fig. 13.2. Detected signal power vs. frequency for fiber lengths of 70.2, 74.6, and 76.7 km. The detected power is relative to the received power at 0 km. (From [126], ©1997 IEEE. Reprinted with permission)

the optical carrier. Fiber chromatic dispersion causes a relative phase shift between these sidebands which causes a detected signal power variation proportional to (12.1) where D is the dispersion parameter, L is the fiber length, and f is the modulation frequency [113, 114]. For multiple carriers, it was shown in [117] that each carrier follows approximately the power variation of (12.1). At 70.2 km, there is a large notch in the spectrum around 28.13 GHz. However, adding lengths of fiber shifts the notch out of the frequency band as shown by the curves at 74.6 and 76.7 km. The signal power variation over the band at 76.7 km is about 10 dB.

Next, a four-tone intermodulation distortion measurement was performed using four unmodulated and random phase mm-wave carriers at 27.96, 27.98, 27.02, and 28.04 GHz. Figure 13.3 shows the detected signal power spectrum at different fiber lengths for an OMD of 24%. The carrier-to-interference ratio (CIR) was measured as the ratio of the lower power at adjacent channel 2 or 4 to the intermod power at 28 GHz. The CIR was 45 dB at a fiber distance of 0 km due to nonlinearities in the upconverter and unlinearized EOM of Fig. 13.1. At distances of 54.5 and 74.6 km, the carrier-to-CTB has a value

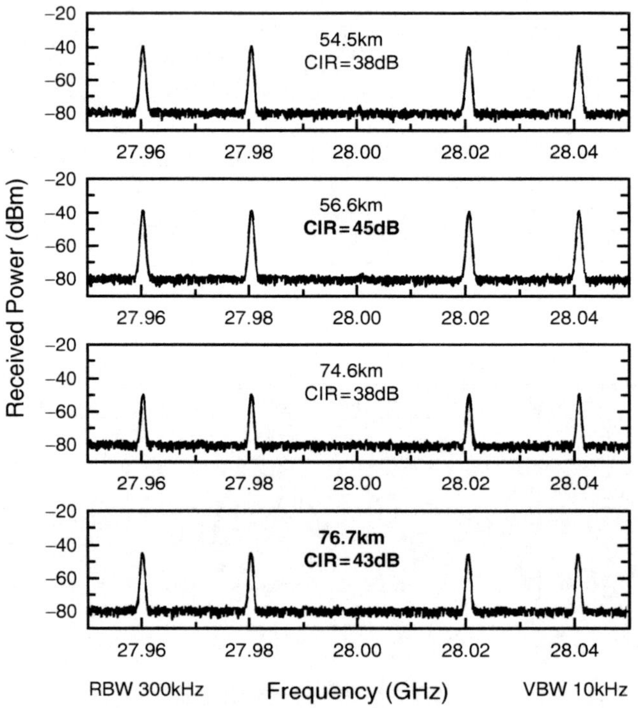

Fig. 13.3. Received signal spectrum at fiber distances of 54.5 km (CIR = 38 dB), 56.6 km (CIR = 45 dB), 74.6 km (CIR = 38 dB), and d) 76.7 km (CIR = 43 dB). (From [126], ©1997 IEEE. Reprinted with permission)

of CIR $=38\,$dB. However, at 56.6 km a CIR $=45\,$dB was measured and at 76.7 km the CIR was 43 dB. This increase in CIR at certain fiber distances agrees with the simulated results reported in [117]. The actual limitation to the CNR at these distances came from the low receiver responsivity and the electrical amplifier thermal noise. The CNR was about 20 dB in a 7-MHz bandwidth at the 76.7-km fiber length. The CNR is given in decibels by

$$\mathrm{CNR} = 10\log_{10}(mRP)^2 25 - kT_0 - NF - 10\log_{10}B \qquad (13.1)$$

where m is the optical modulation depth (0.24), R is the detector responsivity (0.1 A/W), P is the optical power (18 μW), kT_0 is thermal noise ($-204\,$dBW/Hz), NF is the amplifier noise figure (4 dB), and B is the receiver bandwidth (about 7 MHz). The insertion loss from the modulator drive to the detector output is 87 dB, giving an equivalent input noise (EIN) of $-87\,$dBm/Hz.

With a receiver optimized for the 28-GHz band, or with optical amplification, the thermal noise can be overcome. If for example a resonant detector with a higher responsivity of 0.85 A/W [133] was used, the sensitivity should be improved by 9.3 dB. Lower noise amplifiers will also improve the receiver sensitivity. Assuming an erbium-doped fiber amplifier (EDFA) gain of 15 dB and a noise figure of 10 dB, the CNR in the signal-spontaneous beat noise limit is 45 dB [123]. In this case, the CNR is limited by the CIR of 43 dB. These CIR values are well within the requirements for transmission of QPSK modulated carriers. QPSK requires a CNR $=16\,$dB for 10^{-9} symbol error probability, and a CIR $=30\,$dB gives a CNR penalty of less than 0.2 dB [124].

To demonstrate data transmission, a 10 Mb/s PRBS data stream of length $10^{23}-1$ was used to modulate a carrier using the QPSK modulation format. After raised-cosine filtering the signal bandwidth was about 7 MHz. This carrier was upconverted to the center channel at 28 GHz. At this frequency, the intermodulation distortion due to the random phase interfering channels at 27.96, 27.98, 28.02, and 28.04 GHz was largest. The received mm-wave spectrum at 76.7 km is shown in Fig. 13.4.

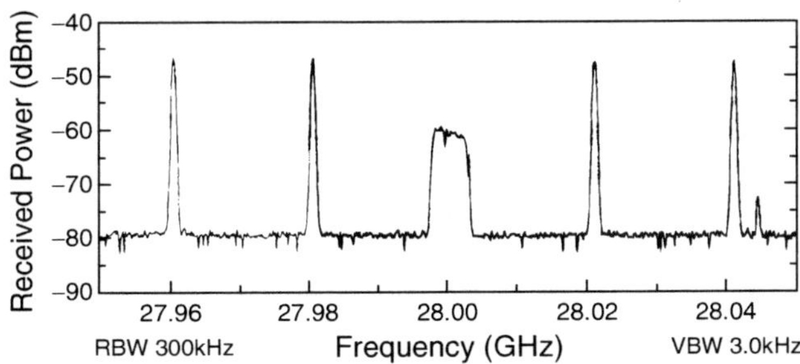

Fig. 13.4. Received mm-wave spectrum after 76.7 km of fiber showing the QPSK modulated carrier and four unmodulated interfering carriers. (From [126], ©1997 IEEE. Reprinted with permission)

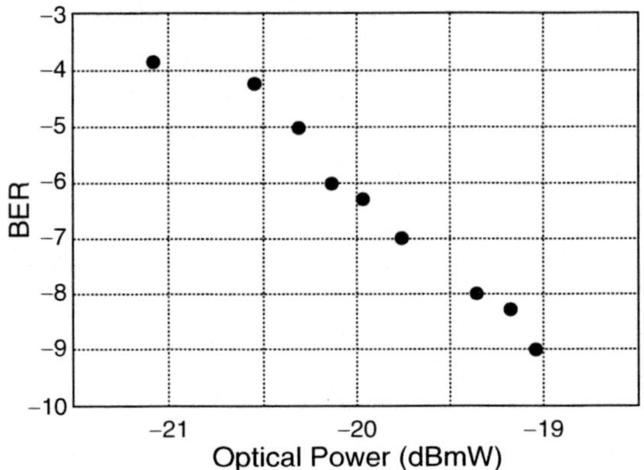

Fig. 13.5. BER of the QPSK channel vs. received optical power at 76.7 km of fiber. (From [126], ©1997 IEEE. Reprinted with permission)

The detected modulated carrier was downconverted and demodulated at the receiver. At the received optical power of 18 mW at a fiber distance of 76.7 km there were no detectable errors in the received data stream. The BER versus optical power at 76.7 km is shown in Fig. 13.5.

This chapter presents distortion measurements for fiber transmission of multichannel mm-wave signals. It is shown that even at fiber distances of up to 77 km the intermodulation distortion is low enough for acceptable transmission of multiple digitally modulated mm-wave carriers. To our knowledge, this is the first demonstration of multichannel mm-wave transmission over long fiber distances. A QPSK modulated carrier at 28.00 GHz was successfully transmitted over 76.7 km of NDSF without optical amplification. In the presence of intermodulation distortion from four adjacent channels bit error rates below 10^{-9} were still achieved. The CNR was limited by the receiver thermal noise. Optical amplification and/or a receiver optimized for mm-wave frequencies can be used to overcome this limit. No electrical linearization was needed on the mm-wave external optical modulator. The significant impact of these results on the mm-wave fiber wireless system architecture is that a 28-GHz optical fiber link can be used to distribute broad-band QPSK information at mm-waves from a headend location to remote hub antenna sites up to at least 77 km away for broadcast services.

13.2 39 GHz Fiber-Wireless Transmission of Broadband Multi-Channel Compressed Digital Video

Figure 13.6 shows two examples of broadband systems which are based on the systems described in [134–136]. For each architecture, a central office or headend services up to 300,000 users. Five to fifteen fiber links connect to

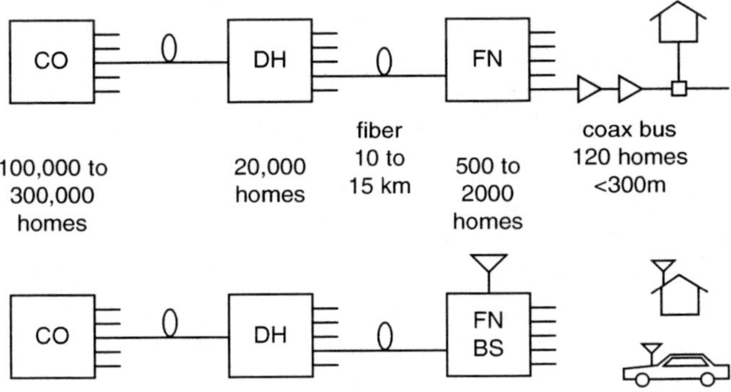

Fig. 13.6. Broadband systems. (From [112], ©1996 IEE. Reprinted with permission)

distribution hubs which service 20,000 users. Each distribution hub uses 10–40 fiber links to distribute signals over up to 15 km to fiber nodes. Each fiber node services an area with 500–2,000 customers.

The two systems differ in the connection from the fiber node to the user for which the service provider has two main choices. Coaxial cable and amplifiers have been proposed for the hybrid fiber-coax system shown in Fig. 13.6 (upper). However, the wireless connection from the fiber node to the user shown in Fig. 13.6 (lower) can be quicker and less expensive for the service provider to install, especially in urban areas and regions with difficult terrain. Another important advantage over a wired network is that the wireless solution enables user mobility.

Because of crowding at lower frequencies, only the mm-wave frequency range offers the bandwidth required for free-space transmission of broadband spectra (1–2 GHz). Given that the remote antenna in Fig. 13.6 (lower) transmits at mm-wave frequencies, there is still a question as to how to deliver signals from the central office to the base station. The broadband signals could be transmitted over fiber at an IF frequency and then mixed-upward to mm-wave frequencies at each base station antenna site.

Alternatively, there are important advantages to using a mm-wave optical transmitter to transmit the mm-wave information over the fiber to each base station:

1. Simplification of base stations: the mm-wave upconverter, phase-locking and control equipment are removed from the base station resulting in significant cost and complexity reduction.

2. Centralized control of mm-wave signals: because the base stations do not change the frequencies of the signals, the mm-wave information sent from the central office will be the same information transmitted through free-space at the remote antenna site. This allows the central office to remotely monitor its transmission frequencies, *which is especially important for compliance with the FCC.*

3. Stabilization of mm-wave signals: because there are many base stations, having a mm-wave upconverter at each antenna site implies it must be inexpensive. A low cost mm-wave oscillator for the upconverter will have significant phase noise and frequency drift. Each base station services 500 to 2,000 users, while each fiber link can service $\sim$30,000 users. Therefore, moving the mm-wave upconverter to the central office substantially reduces the per user cost, where a more expensive but highly stable mm-wave local oscillator (LO) can be used.

For the above reasons mm-wave optical fiber links are desirable for delivery of mm-wave signals to remote antenna sites for broadband wireless networks.

Broadband ($>$500 MHz) multiple channel services have stringent linearity requirements because of their sensitivity to channel interference. For fiber links, the main source of distortion is the broadband optical transmitter. High-speed external optical amplitude modulators (EOM) have been demonstrated at frequencies well into the mm-wave frequency band [120]. However, the optical intensity against modulation amplitude response is nonlinear, which causes intermodulation distortion (IMD).

Another broadband mm-wave optical transmitter is the mm-wave electro-optical upconverter described in [137] which uses a low frequency laser diode cascaded with a high-speed external optical modulator. Because the multiple IF channels modulate a linearized diode, the laser intermodulation distortion should be small. The mm-wave electrical LO drives the EOM and modulates the laser optical output, resulting in an upconversion of the IF channels to mm-wave frequencies. The only distortion products contributed by the EOM are harmonics of the single mm-wave LO frequency which are well beyond the frequency response of the system.

The broadband mm-wave optical transmitter of [137] was used to demonstrate transmission of multichannel digitally compressed MPEG-2 video over a 39 GHz mm-wave fiber-wireless link as shown in Fig. 13.7. The MPEG-2 encoder bits are raised cosine filtered and mapped into one of up to 1,620 MHz-wide QPSK channels which occupy the 500 MHz of bandwidth from 300 to 800 MHz. The laser was a commercially available Ipitek FiberTrunk-900 1,310 nm linearized DFB laser transmitter with a CATV bandwidth of 50–900 MHz which was wide enough for the 500 MHz bandwidth of the input channels. It had optical isolation, temperature cooling and bias control.

The high-speed 50 GHz EOM was a Mach–Zehnder $LiNbO_3$ optical amplitude modulator [120] biased at the half-intensity point. The electro-optical

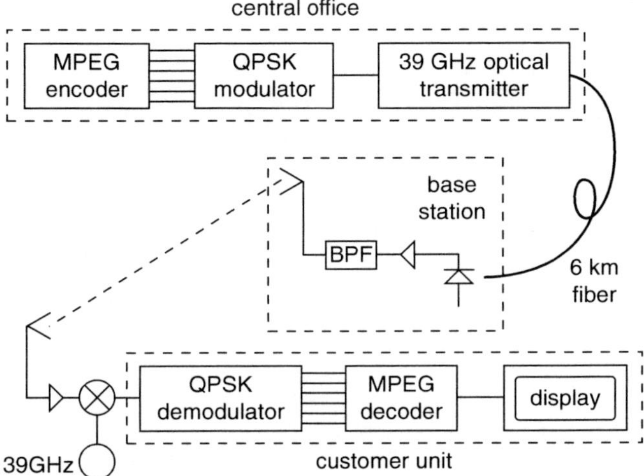

Fig. 13.7. Millimeter-wave fiber-wireless digital video system. (From [112], ©1996 IEE. Reprinted with permission)

upconverter signal bandwidth was limited only by the laser modulation bandwidth which was ∼900 MHz.

The 300–800 MHz frequency band of QPSK channels were used to intensity modulate the CATV laser. The EOM was driven at 39.5 GHz with the HP 83650A Synthesized Sweep Signal Generator. As described in [137], the channels are upconverted to 39 GHz. The EOM modulation depth was ∼0.7.

To demonstrate fiber distribution of the mm-wave signals, 6 km of single-mode fiber was used between the optical transmitter and the base station. There was ∼0.5 mW of optical power at the detector. No optical amplification was needed.

The base station consisted of a high-speed photodiode, a 39 GHz bandpass filter, a high-power mm-wave amplifier, and an antenna. The 500 MHz wide spectrum of channels centered at 39 GHz was amplified to 5 dBm and transmitted through an antenna.

The received and amplified spectrum around 39 GHz is shown in Fig. 13.8. The wireless link loss was 55 dB, which at this frequency corresponds to a free space propagation path of over 1 km when high gain (35 dBi) transmit and receive antennas are used. High gain antennas are used for example in point-to-point wireless links.

At the receiver, the 39 GHz signal was downconverted back to the IF frequencies of 300–800 MHz using a 39.5 GHz LO and a Watkins–Johnson mm-wave mixer. The QPSK demodulator handles automatic gain control, carrier and timing recovery. The MPEG-2 decoder had equalization and the decoded video was observed on a monitor.

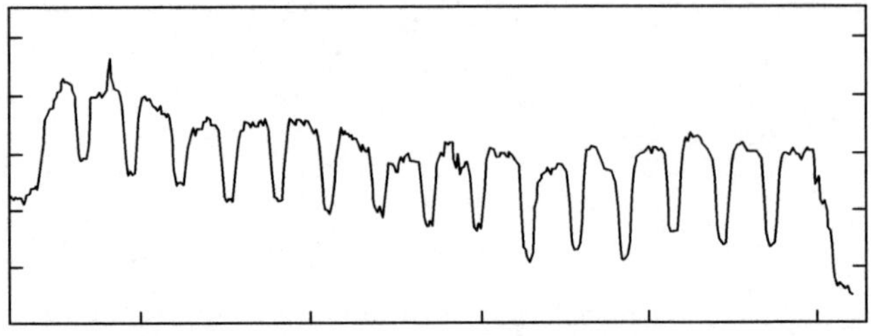

Fig. 13.8. Received spectrum at 39 GHz. (From [112], ©1996 IEE. Reprinted with permission)

Seventy digital video channels were observed using the optical transmitter. There was good video with no decoding errors in any of the video channels.

As a conclusion this chapter describes the importance of broadband mm-wave fiber-optic links for the distribution of mm-wave signals for broadband (>500 MHz) wireless services. A low distortion mm-wave electro-optical up-converter can be used to transmit broadband, multiple-channel digital video over a mm-wave fiber-wireless link using 6 km of optical fiber and an equivalent 1 km wireless point-to-point link.

Application of Linear Fiber Links to Wireless Signal Distribution: A High-Level System Perspective

The use of analog (a.k.a. linear) fiber-optic links as the connecting infrastructure in wireless microcellular networks has been proposed [138–142]. Wireless systems must provide uniform radio coverage to spatially distributed mobile users in a cost effective manner. Small (radius $\sim 300\,\mathrm{m}$) radio microcells can serve a high density of users, and require low user handset transmit power compared to large ($r \sim 1\,\mathrm{km}$) macrocells in existing systems. A microcellular network can be implemented by using a fiber-fed distributed antenna network as shown schematically in Fig. 14.1. The received RF signals at each antenna are transmitted over an analog fiber-optic link to a central base station where all multiplexing/demultiplexing and signal processing are done. In this way, each remote antenna site simply consists of the antenna, amplifier, and a linear (analog) optical transmitter. The cost of the microcell antenna sites can thus be greatly reduced therefore rendering deployment of these networks practical. The required dynamic range of the analog optical transmitter is a major factor in cost. Previous analysis [138] on dynamic range requirements assumed an absolute spur-free condition for each FDM channel, and resulted in a link dynamic range requirement of $>100\,\mathrm{dB}$ ($1\,\mathrm{Hz}$). This chapter investigates the dependence of this dynamic range requirement in realistic scenarios, depending on the number of active voice channels, the density of antenna coverage, and network protocol. For a *single* antenna serving a cell, in real life traffic call-blocking occurs due to the random nature of initiation/departure of callers served by a given base station, It is not necessary for fiber-optic links which serve this type of network to have performances exceeding limitations imposed by fundamental traffic considerations, in this chapter it will be shown that by conforming the performance of the fiber links to the small ($<0.5\%$) call-blocking probability, commonly encountered in busy wireless traffic. A modest dynamic range of $91\,\mathrm{dB}$ is acceptable for serving 20 FDM voice channels in the cell. Furthermore, it is shown that by using multiple fiber-fed antennae per cell – now designated a "microcell," and employing an appropriate network protocol, the required dynamic range of

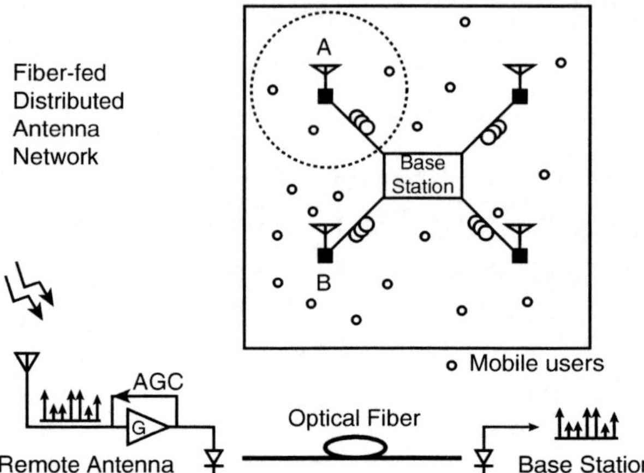

Fig. 14.1. Fiber-fed distributed antenna wireless network using analog (a.k.a. linear) links (fiber or otherwise)

the links can be further reduced to a modest 73 dB and still suffices for an acceptable wireless network performance.

The dynamic range of fiber-optic links is commonly specified by the range of input powers over which the output signal is "spur-free" for a two-tone input (see Sect. A.1). In a microcellular wireless system, this two-tone specification over-estimates the required link performance since it corresponds to the rare, unfortunate situation when two high power mobile emitters are assigned to adjacent frequency channels. To determine practical values for the required dynamic range of optical links (or any other type of RF cable links, for that matter), a statistical simulation of user access in a wireless microcell is described in this chapter. The model is based on the AMPS cellular system which uses FDM (Frequency Division Multiplex) for multiple access, and which requires an 18 dB carrier-to-interference (C/I) ratio, with an allocation of 30 kHz of bandwidth per voice channel. A standard model for multipath environments is employed [143], in which the received RF power varies as $(1/d)^4$ where d is the distance from the antenna to the user. Experimental measurements in a typical line-of-sight urban microcellular system [144] have demonstrated that this model empirically provides a lower bound for the received power including local fading effects of the signal. It is also assumed that the users are only allowed to within 5 m of each antenna site (as in the case of an indoor cell where antennae are mounted above the drop ceiling), which leads to a maximum RF power variation of $\sim$70 dB for a 300-m microcell. Although the effects of shadowing are not included in our model, existing mobile handsets typically have some limited power control capability which can compensate for shadowed environments. A series of simulations are performed in which users appear randomly in the cell, and the spectrum of received RF powers at each antenna is amplified by an automatic gain control

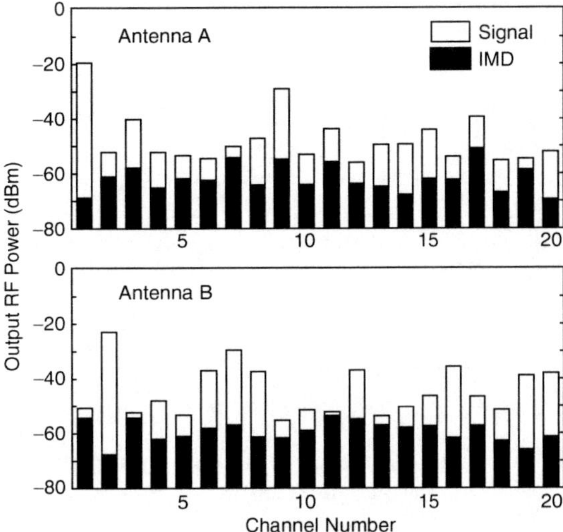

Fig. 14.2. Typical spectrum of received signal and IMD powers at antennas A and B shown in Fig. 14.1. For this plot, the transmitter dynamic range is 70 dB (1 Hz)

(AGC) amplifier that maintains the weakest signal to 18 dB above the noise floor of the link. Next, the intermodulation distortion (IMD) terms from the optical transmitter at each channel are calculated as:

$$\mathrm{IMD} = \gamma \left(\sum_{2\omega_i - \omega_j} P_i^2 P_j + \sum_{\omega_i + \omega_j - \omega_k} 4 P_i P_j P_k \right) \tag{14.1}$$

where the summations are over all combinations of ω_i, ω_j, and ω_k that fall on the given channel and the proportionality constant is determined by the two-tone dynamic range of the link. Figure 14.2 shows a typical spectrum of output signal powers and the resulting intermodulation distortion (IMD) terms from two of the antennas shown in Fig. 14.1. Upon transmission through the link which connects the antenna to the base station, any channel that does not meet the minimum 18-dB C/I ratio is counted as a blocked call. This process is repeated until the product of the number of runs and the number of voice channels equals 10^5. The average percentage of blocked calls is then calculated as a function of the link two-tone dynamic range.

First, a hypothetical circular microcell ($r = 300$ m) is considered with only one fiber-fed antenna covering the cell. Figure 14.3 shows the average blocking probability as a function of the link twotone dynamic range for 5, 10, and 20 available FDM voice channels. As expected, the number of blocked calls decreases with increasing link performance. The dashed line in the figure corresponds to a relatively small blocking probability of 0.5% (an acceptable standard in cellular telephony). The link performance required to achieve this blocking level increases with the number of channels due to the increase in the

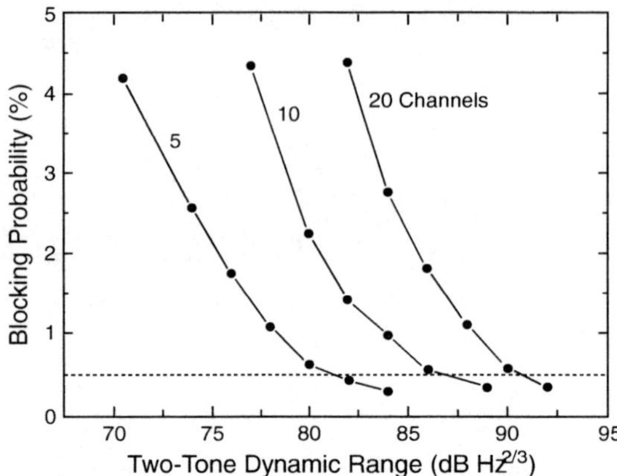

Fig. 14.3. Percentage of blocked calls as a function of transmitter dynamic range for a single fiber-fed 300-m-radius microcell for 5, 10, and 20 voice channels

number of intermodulation products. For 20 channels, the criterion is met for a modest link dynamic range of 91 dB. Next, the scenario of multiple antennas within an 1,800 m^2 cell is considered. When using these multiple antennas, the base station must decide how to assign users to a particular antenna. For example, referring to Fig. 14.2, channel 1 has a higher C/I at antenna A while channel 2 has a higher C/I at antenna B. Two different protocols are possible for processing the signals from the multiple antennas. First, consider the case that each channel is assigned to a particular antenna based on the strength of its received signal power. Second, each channel is assigned based on the C/I of that channel. This C/I protocol is more difficult to implement than the maximum power protocol since it requires a mean to measure the quality of the received signals on each channel. Figure 14.4 shows the blocking probability with 4 and 9 antennas per cell for both of these protocols. Notice the dramatic influence of the network protocol on the optical transmitter requirement. For the maximum signal protocol, there is not a significant difference between using 4 and 9 antennas, and the link dynamic range requirement remains in the ~90-dB range. This can be understood by realizing that the maximum signal protocol is equivalent to dividing the cell into several smaller cells. In these smaller cells, the variation in received RF power is reduced; however, the probability that a user will saturate an antenna is increased since there are more antennas in the cell. In order to take advantage of the spatial diversity of the distributed antenna network, the maximum C/I protocol must be implemented. These simulations indicate that the required dynamic range is reduced to 78 dB for 4 antennas and to 73 dB for the 9 antenna case. With this protocol implemented, when a "close-in" user monopolizes a particular antenna, the other users in the vicinity of that antenna can be "picked up" by one of the other antennas that is not saturated. In this case, the network performance is very poor only for the rare, unfortunate situation when *all* of

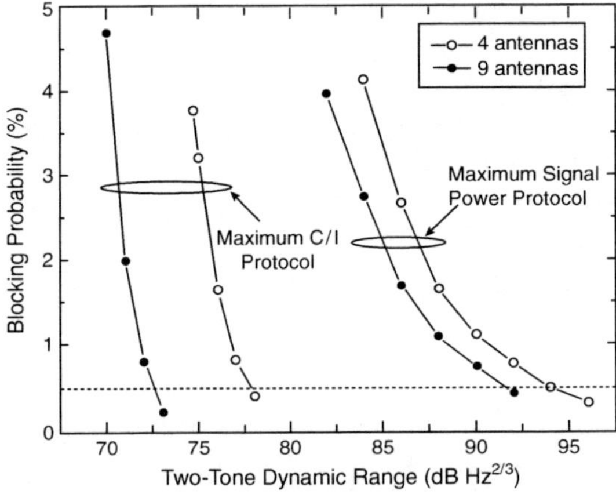

Fig. 14.4. Percentage of blocked calls as a function of transmitter dynamic range for 4 and 9 antennas and two different signal selection protocols. This simulation considers an 1,800-m-square area with 20 available voice channels

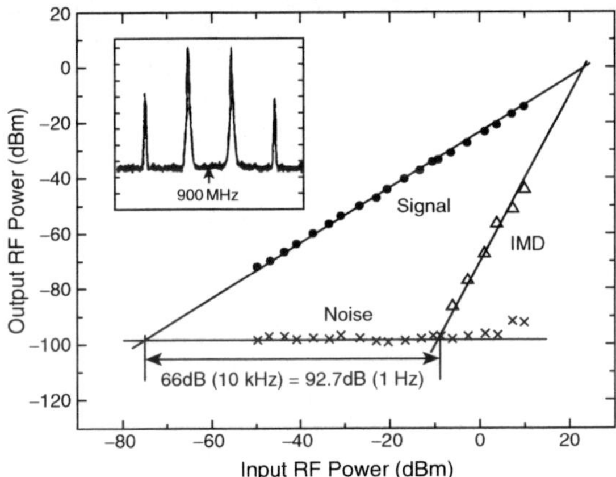

Fig. 14.5. Measured two-tone dynamic range of a self-pulsating CD laser at 900 MHz. The inset (10 dB/div) shows the two tones centered at 900 MHz and the induced intermodulation products

the distributed antennas in the cell are saturated by close-in users. Even further reductions in the required link dynamic range can be obtained by taking advantage of dynamic channel assignment (DCA) strategies that allocate user channels such that the generated intermodulation products discussed above are minimized.

To put in perspective the modesty of these dynamic range requirements determined from the above analysis, Fig. 14.5 shows the experimentally measured two-tone dynamic range of an inexpensive Sharp compact disk (CD) laser. Operating at an optical power of 6.9 mW without an optical isolator, the dynamic range at 900 MHz is 92.7 dB. The inset shows the two tones centered at 900 MHz and the induced intermodulation products. This measurement may have significant implication in the use of very low cost optical transmitters in the fiber-fed microcellular networks discussed above.

In conclusion, the dynamic range requirements of optical transmitters used in fiber-fed distributed antenna networks are determined using a statistical model for the number of dropped calls due to the generated IMD products. A single antenna microcell fed with a 91 dB link can support 20 FDM channels with a dropped call probability of 0.5%. By covering a cell with multiple antennas, and implementing an optimum protocol at the base station, the link linearity requirement can be reduced to <80 dB. These results may have significant implications in the practical implementation of next-generation microcell personal communication networks.

15

Improvements in Baseband Fiber Optic Transmission by Superposition of High Frequency Microwave Modulation

15.1 Introduction

It is well known that phase noise fluctuations in the output of a semiconductor laser can produce intensity noise fluctuations upon transmission through a fiber-optic link due to interferometric phase-to-intensity conversion [145–148]. In a single-mode fiber link, the interferometric conversion occurs when multiple reflections occur between a pair of fiber interfaces (Fig. 15.1). Even in the absence of such fiber discontinuities, Rayleigh backscattering in a sufficiently long piece of fiber can cause similar effects [149]. If a laser source is used in a multimode fiber link, the different transverse modes of the fiber interfere with one another and produce the well known "modal noise." This chapter specifically studies the former case (multiple discrete reflections in a single-mode fiber link) although it is straight forward to extend the formalism to the case of modal noise in multimode fibers. The nature of interferometric noise has been studied in [145–147]. It was shown that this excess noise can cause bit-error-rate floors [69], and the system performance has been evaluated as a function of the number and/or amplitude of the reflections [150].

To the extent that such interferometric noises arise from interference of the laser output with a delayed version of itself, it is obvious that reduction of laser coherence can eliminate these noises. Indeed, there have been proposals and early demonstrations that by applying a high frequency modulation to the laser, the coherence of the laser can be reduced, leading to a reduction of the type of interferometric noise mentioned above. However, except in the extreme case of a very deep modulation (where the laser output is almost pulse-like and each pulse is incoherent with the previous ones), one does not expect the laser output to be rendered totally incoherent by the applied modulation, but instead, the lasing wavelength "chirps" sinusoidally at the modulation frequency. Can one expect interferometric noises to be reduced (or even totally eliminated) under this situation? The following analysis will show that the answer is positive, provided that proper choices of modulation format and parameters are employed.

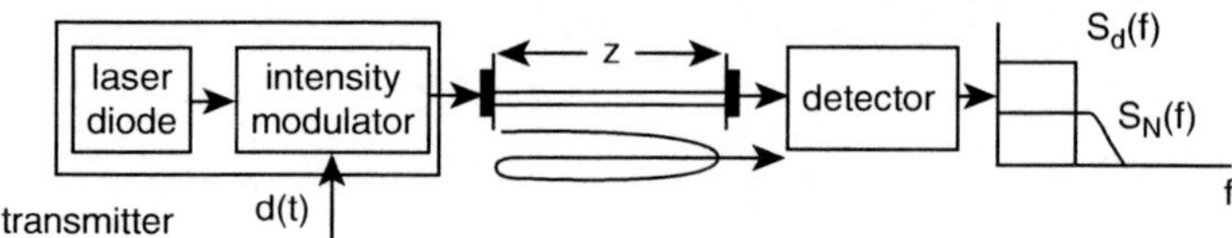

Fig. 15.1. Fiber-optic system exhibiting multiple reflections between a pair of fiber interfaces

The nature of interferometric noise is first described in Sect. 15.2 [145–147]. Next, a number of high frequency superimposed modulation formats are considered. The order of the sequence of these expositions is chosen to best illustrate the nature of the mechanisms responsible for the interferometric noise reduction.

15.2 Interferometric Noise

The results derived in this section are based on [145–147].

Consider the intensity noise generated in a single-mode (SM) fiber optic link through interferometric FM-AM conversion due to, for example, double reflection between a pair of retro-reflecting optical surfaces, such as an imperfect fiber optic connector (Fig. 15.1). The laser is assumed to be of single-frequency (wavelength), and it is also assumed that the data is intensity modulated onto the optical output from the laser using an *external chirp-free modulator*. The electric field at the input to the fiber is given by

$$E(t) = \sqrt{P(t)}e^{j\Omega_0 t + \phi(t)} \tag{15.1}$$

where $P(t)$ is the (externally) modulated optical power from the laser, Ω_0 is the optical carrier frequency and $\phi(t)$ is the laser phase noise. In the situation where there is a pair of retro-reflective discontinuities as shown in Fig. 15.1, the field at the output of the fiber is represented by the superposition of the original input field with delayed version of itself

$$E_{\text{out}}(t) = \psi_1 E(t - t_1) + \psi_2 E(t - t_2) \tag{15.2}$$

where ψ_1 and ψ_2 are the relative field intensities, and t_1 and t_2 are delays. Without loss of generality, one can assume that $\psi_1 = 1$, $\psi_2 = \psi \ll 1$, $t_1 = 0$, and $t_2 = \tau$.

The modulated optical *power* from the (externally modulated chirp-free) laser transmitter can be written as

$$P(t) = P_0 d(t) \tag{15.3}$$

where P_0 is the average output laser power (assume a lossless external modulator) and $d(t)$ represents band limited data. Intrinsic laser intensity noise is neglected in the following analysis, since it is much smaller than the other noises being considered.

The electric field at the fiber output is

$$E_{\text{out}}(t) = \sqrt{P_0 d(t)}\,e^{j\Omega_0 t}e^{j\phi(t)} + \psi\sqrt{P_0 d(t-\tau)}\,e^{j\Omega_0(t-\tau)+j\phi(t-\tau)} \qquad (15.4)$$

Then, the *optical intensity* at the output of the fiber is

$$i(t) = |E_{\text{out}}(t)|^2 + \psi^2 |E(t-\tau)|^2 + 2\psi\Re\{E(t)e^*(t-\tau)\} \qquad (15.5)$$

where $\Re\{...\}$ denotes the real part of the quantity. One can identify the signal $(i_S(t))$ and the noise part $(i_N(t))$ of $i(t)$ as

$$i_S(t) = P_0[d(t) + \psi^2 d(t-\tau)] \approx P_0 d(t) \qquad (15.6)$$

where it was assumed $\psi \ll 1$ and

$$i_N(t) = 2\psi P_0 \sqrt{d(t)d(t-\tau)}\,\cos[\Omega_0\tau + \phi(t) - \phi(t-\tau)] \qquad (15.7)$$

The laser phase noise $\phi(t)$ is modeled to follow Gaussian probability density function and $\phi(t)$ and $\phi(t-\tau)$ are correlated in such a way that [145]

$$\left\langle (\phi(t) - \phi(t-\tau))^2 \right\rangle = \frac{|\tau|}{\tau_c} \qquad (15.8)$$

where τ_c is the laser coherence time, and $\langle\rangle$ denotes statistical averaging. In this case, (15.8) corresponds to a Lorentzian lineshape of the spectral distribution. The expression (15.7) is a *general* result, which will be used to derive the noise spectrum in several cases. With the above assumptions, the fine structure of the spectrum due to relaxation oscillations [151, 152] is ignored because it is of secondary importance in this analysis.

It is then straightforward to find the signal and the noise spectrum. The task is to find these corresponding autocorrelation functions. They are:

$$\begin{aligned}
R_S(\delta t) &= E\{i_S(z,t)i_S(z,t+\delta\tau)\} \\
&= (1+\psi^2)P_0^2 R_d(\delta\tau) \qquad (15.9) \\
R_N(\delta t) &= E\{i_N(z,t)i_N(z,t+\delta\tau)\} \\
&= 2\psi^2 P_0^2 R_{dd}(\delta\tau)[R_-(\delta\tau) + R_+(\delta\tau)\cos(\Omega_0\tau)] \qquad (15.10)
\end{aligned}$$

where $R_d(\delta\tau)$ is the autocorrelation function of $d(t)$, and

$$R_{dd}(\delta\tau) = E\left\{ \sqrt{d(t)d(t+\delta\tau)d(t-\tau)d(t+\delta\tau-\tau)} \right\} \qquad (15.11)$$

The corresponding power spectral densities are denoted by $S_d(f)$ and $S_{dd}(f)$ and can be obtained from Fourier transforming R_d and R_{dd}. To compute $R_{dd}(\delta\tau)$ one needs to specify the data statistics. However, for the purpose here the explicit knowledge of $R_{dd}(\delta\tau)$ is not needed, one only needs to specify the bandwidth of the Fourier transform of R_{dd}. If one assumes that $d(t)$ consists of ideal rectangular pulses, then the bandwidth of $S_{dd}(f)$ is equal to the bandwidth of $S_d(f)$.

The expression in [ ] in (15.10) is recognized as the interferometrically converted laser phase-to-intensity noise of a CW laser in the absence of data modulation. The expressions R_+ and R_- are given by

$$R_-(\delta\tau) = \langle\cos[\phi(t) - \phi(t + \tau) - \phi(t + \delta\tau) + \phi(t + \delta\tau + \tau)]\rangle \quad (15.12)$$
$$R_+(\delta\tau) = \langle\cos[\phi(t) - \phi(t + \tau) + \phi(t + \delta\tau) - \phi(t + \delta\tau + \tau)]\rangle \quad (15.13)$$

and have been previously calculated [145, 146]. The variations of the term $R_+(\delta\tau)\cos(\Omega_0\tau)$ are of the order of the laser wavelength. One should be interested in the macroscopic variations, which are on much bigger scale than those due to the term involving $R_+(\delta\tau)$. For this reason the term including $R_+(\delta\tau)$ will be neglected. $R_-(\delta\tau)$ is given by [145]

$$R_-(\delta\tau) = \exp\left[-\frac{1}{2\tau_c}(2|\tau| - |\tau - \delta\tau| - |\tau + \delta\tau| + 2|\delta\tau|)\right] \quad (15.14)$$

For sufficiently long τ, the corresponding power spectral density of $R_-(\delta\tau)$ (denoted by $S_-(f)$), given by the Fourier transform of $R_-(\delta\tau)$ assumes the Lorentzian lineshape of the lasing field down-converted to baseband, with a typical width of 10–100 MHz.

Then, the noise autocorrelation function becomes

$$R_N(\delta\tau) = 2\psi^2 P_0 R_{dd}(\delta\tau) R_-(\delta\tau) \quad (15.15)$$

Its power spectral density is given by

$$S_N(f) = 2\psi^2 P_0 S_{dd}(f) * S_-(f) \quad (15.16)$$

where the "$S(f)$'s" are the Fourier Transforms of the "$R(\delta\tau)$'s", and where $*$ denotes convolution.

The power spectral densities are schematically illustrated in Fig. 15.2. Note that the S/N ratio cannot be increased by increasing the data signal power, since the noise power also increases correspondingly as per (15.16). It illustrates clearly the deleterious effect of interferometric FM-IM noise on imposing an upper limit on the S/N ratio of the transmitted data, regardless of signal power.

15.2.1 Superimposed High-Frequency Modulation: External Phase Modulation

Consider the case in Fig. 15.3b which is similar to Fig. 15.3a but with an external phase modulator placed at the output of the laser. The modulator

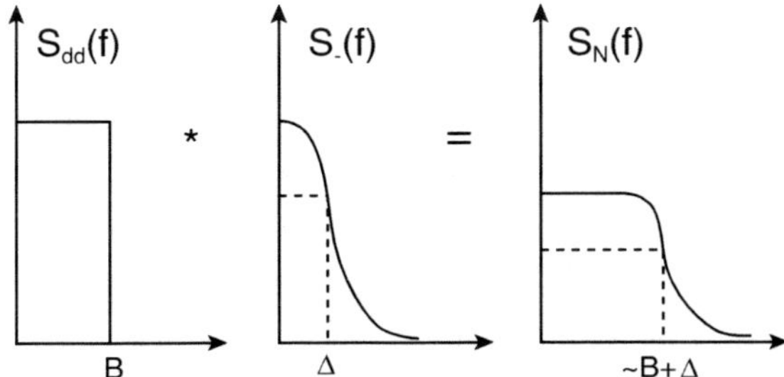

Fig. 15.2. Schematic illustrations of power spectral density functions $S_{dd}(f)$, $S_-(f)$, and $S_N(f)$. It is assumed that the modulation data is band-limited, and therefore so is $S_{dd}(f)$

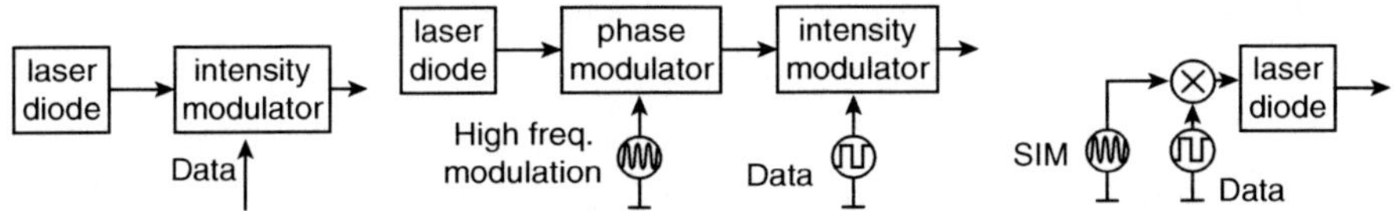

Fig. 15.3. High frequency modulation formats: (**a**) no superimposed high frequency phase modulation; (**b**) superimposed external phase modulation; (**c**) directly superimposed high frequency modulation (SIM)

is driven by a single RF tone at a "high" frequency (higher than the data bandwidth, to be specified later). The optical field at the input to the fiber is then

$$E(t) = \sqrt{d(t)P_0}\,e^{j\Omega_0 t}e^{j\phi(t)}e^{ja\cos(\omega_0 t)} \tag{15.17}$$

where a is the phase modulation index and $f_0 = \omega_0/2\pi$ is the RF modulation frequency. Following a similar procedure as in the last section, the output signal intensity is

$$i_S(z,t) = d(t)P_0 \tag{15.18}$$

where $d(t)$ and P_0 are the same as before, and the noise intensity at the fiber output is, according to (15.7)

$$i_N(z,t) = 2\psi\Re\{E(t)e^*(t-\tau)\} \tag{15.19}$$

$$= 2\psi P_0\sqrt{d(t)d(t-\tau)}\cos\left[\Omega_0\tau + \Delta\phi(t,\tau) + A\sin\left(\omega_0 t - \frac{\omega_0\tau}{2}\right)\right]$$

where

$$A = -2a\sin\left(\frac{\omega_0\tau}{2}\right) \tag{15.20}$$

and $\Delta\phi(t,\tau) = \phi(t) - \phi(t-\tau)$. The noise power spectral density is the Fourier transform of the noise autocorrelation function $R_N(t,\delta\tau)$. In this case it will be necessary to perform both time and statistical averaging to properly model the nonstationary conditions [153].

The autocorrelation function of the noise term $i_N(z,t)$ is given by

$$
\begin{aligned}
R_N(t,\delta\tau) &= \langle i_N(z,t)i_N(z,t+\delta\tau)\rangle \\
&= 2\psi^2 P_0^2 R_{dd}(\delta\tau)R_-(\delta\tau) \\
&\quad \cdot \cos\left[2A\sin\left(\frac{\omega_0\delta\tau}{2}\right)\cos\left(\omega_0 t - \frac{\omega_0\tau}{2} + \frac{\omega_0\delta\tau}{2}\right)\right]
\end{aligned} \quad (15.21)
$$

In deriving (15.21) the $R_+(\delta\tau)$ term is neglected as before. Also, there is a cross term $E\{i_S(t+\delta\tau)i_N(t)\}$ which can be shown to be proportional to $\cos(\Omega_0\tau)$. Using the same arguments as before this term can be neglected. Then, the desired expression for $R_{N_1}(\delta\tau)$, which is obtained by time averaging of $R_N(t,\delta\tau)$ can be written in the following form using Bessel functions expansion:

$$
R_{N_1}(\delta\tau) = \overline{R_N(t,\delta\tau)} \quad (15.22)
$$

$$
= 2\psi^2 P_0^2 R_{dd}(\delta\tau)R_-(\delta\tau)\left[J_0^2(A) + 2\sum_{n=1}^{\infty} J_n^2(A)\cos(n\omega_0\delta\tau)\right]
$$

The noise power spectral density is

$$
S_{N_1}(f) = J_0^2(A)S_N(f) + \left\{\sum_{n=1}^{\infty} J_n^2(A)[S_N(f-nf_0) + S(f+nf_0)]\right\} \quad (15.23)
$$

where $S_N(f)$ is the noise spectrum without superimposed modulation given with (15.16). The expression for $S_{N_1}(f)$ is the primary result of this analysis. It shows that the noise is distributed among the various harmonics of the externally applied phase modulation (Fig. 15.4). Note that in the absence of high-frequency phase modulation, $A = 0 - (15.23)$ simply reduces to that of the intrinsic interferometric noise $-$ $S_N(f)$. With phase modulation, the noise is distributed and transposed to multiples of the superimposed phase

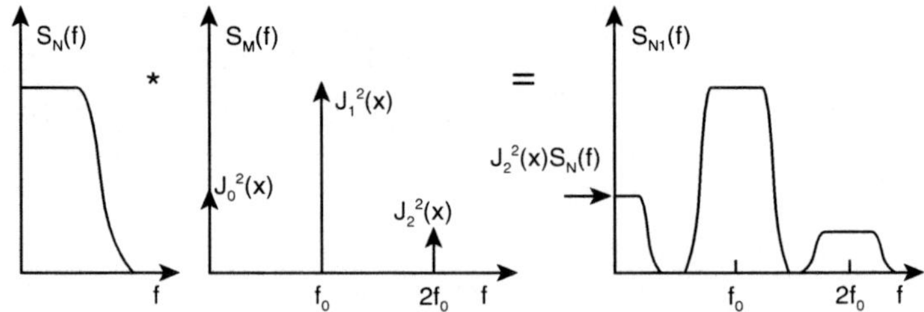

Fig. 15.4. Distribution of the noise power among the different harmonics. Notice how the noise power is transferred from baseband to the higher frequencies where it can be easily filtered

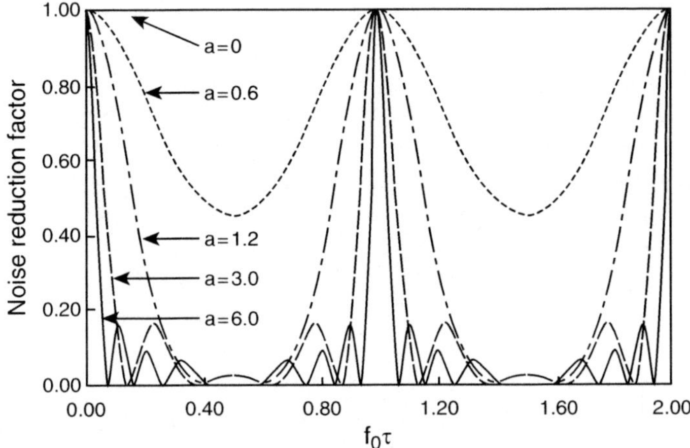

Fig. 15.5. Noise reduction factor vs. $f_0\tau$ for external phase modulation. The phase modulation index a is parameter. The NRF is periodic with period 1

modulation frequency, with "transposition factors" equal to J_n where J_n is the n-th order Bessel function of the first kind.

The part of the noise spectrum of interest is that near baseband, which is given by $J_0(A)^2 S_N(f)$. This expression is valid if the phase modulation frequency is higher than at least twice the bandwidth of the interferometric noise spectrum $S_N(f)$, the latter of which is assumed to be band-limited. The baseband noise is thus reduced by a factor $J_0(A)^2$, which is defined here as the Noise Reduction Factor (NRF). Figure 15.5 shows a plot of NRF *vs* normalized modulation frequency $f_0\tau$ of the superimposed phase. It is obvious that NRF is a periodic function of $f_0\tau$ and the maximum reduction depends on the phase modulation index a. It implies that the NRF is also function of propagation distance down the fiber ($\tau = \text{const.} \times z$).

It is seen from Fig. 15.5 that for very strong phase modulation, the interferometric noise can by-and-large be eliminated except for the unfortunate situation when $f_0\tau = k$, where k is integer. This is somewhat undesirable since a transmitter incorporating a phase modulator at one frequency may work for a particular fiber link, but may not work for another one. In Sect. 15.2.3 it will be shown this situation can be remedied if the applied phase modulation consists *not* of a single high frequency tone but is rendered "noisy" instead, such that its spectrum centers at a high frequency and with a width much larger than $1/\tau$.

15.2.2 Directly Modulated Laser Diode

The effectiveness of a superimposed high frequency phase modulation in reducing interferometrically generated noise has been shown in the above section. However, the arrangement shown on Fig. 15.3b where both external phase and intensity modulators are employed is not desirable in practice. It is far more preferable to apply both the data and high frequency modulation directly to the laser, as shown in Fig. 15.3c. Note that the data and the high frequency signal are first mixed (multiplied) together before being applied to modulating the laser current. In this section, it will be shown that apart from slight quantitative differences, this scheme achieves a similar noise reduction effect as in the ideal case considered previously. For the analysis, one can use the fact that when directly modulating a laser diode, the phase and amplitude of the output lasing field are related by [60]

$$\dot{\phi}_m = \frac{\alpha}{2\pi} \frac{1}{P(t)} \frac{\partial P(t)}{\partial t} \tag{15.24}$$

One assumes that the high frequency modulation current applied to the laser diode is sinusoidal, at frequency f_0. Using the result of a large signal analysis [154], the output intensity is given by

$$P_m(t) = \frac{P_0}{I_0(a)} e^{a\,\cos(\omega_0 t)} \tag{15.25}$$

where $I_0(a)$ is the modified Bessel function of zero order and a is a parameter describing the modulation depth parameter which depends both on the frequency and the modulation amplitude [154]. The phase modulation is given by (15.24) and (15.25) which is, assuming $a = 2\pi$, $\phi_m(t) = a\cos(\omega_0 t)$. The field at the laser output is

$$E(t) = \sqrt{P(t)}e^{ja\,\cos(\omega_0 t)} \tag{15.26}$$

where $P(t) = P_m(t)d(t)$ and one has assumed that the phase modulation due to the data input is negligible compared to that due to the high frequency superimposed modulation. This is justified by the fact that the latter is applied at a much higher frequency than the former and the phase modulation frequency is to first order proportional to the modulation frequency, as evident from the relationship in (15.24). The influence of laser phase noise was neglected. Using the previous results, one can easily derive the expression for the signal and noise intensity at the output of the fiber

$$i_S(z,t) = d(t)P_m(t) + \psi^2 d(t-\tau)P_m(t-\tau) \approx d(t)P_m(t) \tag{15.27}$$

$$i_N(z,t) = 2\psi\sqrt{d(t)d(t-\tau)}\sqrt{P_m(t)P_m(t-\tau)}$$

$$\cdot \cos\left[\Omega_0\tau + \Delta\phi(t,\tau) + A\sin\left(\omega_0 t - \frac{\omega_0\tau}{2}\right)\right] \tag{15.28}$$

The autocorrelation function of the signal is

$$R_S(\delta\tau) = R_d(\delta\tau)\overline{P_m(t)P_m(t+\delta\tau)} \tag{15.29}$$

It can be shown, after some algebra, that the autocorrelation of the noise is

$$\langle i_N(z,t)i_N(z,t+\delta\tau)\rangle = 2\psi^2 R_{dd}(\delta\tau)$$
$$\times \sqrt{P_m(t)P_m(t-\tau)P_m(t+\delta\tau)P_m(t+\delta\tau-\tau)}$$
$$\times R_-(\delta\tau)$$
$$\times \cos\left[2A\sin\left(\frac{\omega_0\delta\tau}{2}\right)\cos\left(\omega_0 t - \frac{\omega_0 t}{2} + \frac{\omega_0\delta\tau}{2}\right)\right]$$
$$(15.30)$$

The term involving $R_+(\delta\tau)$ was neglected as before. If the discussion is limited only to the small signal case, one can write:

$$\sqrt{P_m(t)P_m(t-\tau)P_m(t+\delta\tau)P_m(t+\delta\tau-\tau)}$$
$$= \frac{P_0^2}{I_0^2(a)}\left[I_0(B) + I_1(B)\cos\left(\omega_0 t - \frac{\omega_0\tau}{2} + \frac{\omega_0\delta\tau}{2}\right)\right] \qquad (15.31)$$

where $B = 2a\cos(\frac{\omega_0\tau}{2})\cos(\frac{\omega_0\delta\tau}{2})$. After time averaging, one can get for the autocorrelation function:

$$R_{N_2}(\delta\tau) = \overline{\langle i_N(z,t)i_N(z,t+\delta\tau)\rangle}$$
$$= 2\frac{(\psi P_0)^2}{R_{dd}(\delta\tau)R_-(\delta\tau)I_0(B)}J_0\left[2A\sin\left(\frac{\omega_0\delta\tau}{2}\right)\right] \qquad (15.32)$$

This function is periodic, with period $T = 1/f_0$, where f_0 is the modulation frequency. Again, the noise spectrum has the same form, as in the case of external high frequency modulation. The only term that is of interest is (as in the previous cases) the baseband term. If Fourier series expansion of Bessel functions is used [155], one can get for the baseband term

$$R_{N_0}(\delta\tau) = 2\frac{(\psi P_0)^2}{I_0^2(a)}R_{dd}(\delta\tau)R_-(\delta\tau)$$
$$\times\left[I_0^2\left(a\cos\left(\frac{\omega_0\tau}{2}\right)\right)J_0^2\left(2a\sin\left(\frac{\omega_0\tau}{2}\right)\right)\right.$$
$$\left. + 2I_1^2\left(a\cos\left(\frac{\omega_0\tau}{2}\right)\right)J_1^2\left(2a\sin\left(\frac{\omega_0\tau}{2}\right)\right)\right] \qquad (15.33)$$

Figure 15.6 shows plot of the NRF vs. normalized frequency $f_0\tau$. The results are very similar to the previous cases (Fig. 15.5).

15.2.3 Superimposed Modulation with Band-Pass Gaussian Noise

It was seen in Sects. 15.2.1 and 15.2.2 that for a single tone high frequency phase modulation, the interferometric noise can be reduced except for the unfortunate situations when the modulation frequency and the round trip

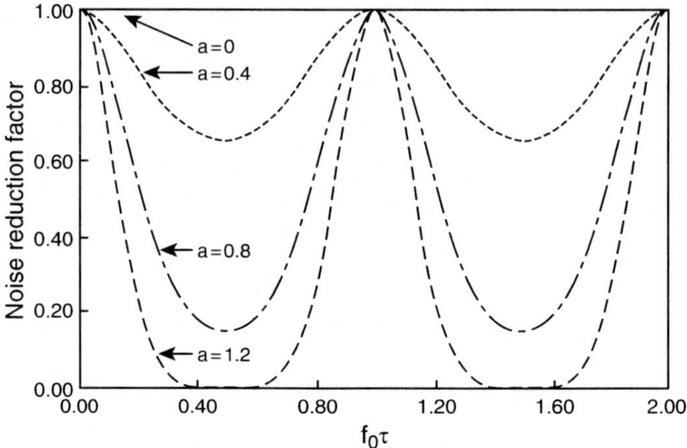

Fig. 15.6. Noise reduction factor vs. $f_0\tau$ for directly modulated LD. a is parameter. The results are very close to those on Fig. 15.5

delay is related by $f_0\tau = k$, where k is integer. To achieve suppression of interferometric noise under all situations independent of τ, one can consider broadening the single tone modulation into a noise band, centered at an arbitrary high frequency. This noise can be generated, for example, by ordinary diode.

To analyze the situation, consider the ideal case where phase modulator is applied externally to the laser output (Fig. 15.3b). Let the modulation applied to the phase modulator of Fig. 15.3b be bandpass noise $n(t)\cos(\omega_0 t)$. The electric field at the input of the fiber is

$$E(t) = \sqrt{d(t)P_0}\,e^{j\Omega_0 t}e^{j\phi(t)}e^{jan(t)\cos(\omega_0 t)} \tag{15.34}$$

Then, following the procedure described in Sect. 15.2.1 one can get for the autocorrelation function of the noise R_{N_3}

$$R_{N_3}(\delta\tau) = 2(\psi P_0)^2 R_{dd}(\delta\tau) R_-(\delta\tau)\exp\{-a^2[2R_n(0) - 2R_n(\delta\tau)\cos(\omega_0\delta\tau)]\}$$

$$\times \exp\{-a^2[-2R_n(\tau)\cos(\omega_0\tau) + R_n(\tau+\delta\tau)\cos(\omega_0(\tau+\delta\tau))$$

$$+R_n(\tau-\delta\tau)\cos(\omega_0(\tau-\delta\tau))]\} \tag{15.35}$$

where $R_N(\delta\tau)\cos(\omega_0\delta\tau)$ the autocorrelation function of the noise at the input of the phase modulator. In principle, the noise power spectral density can then be computed from the autocorrelation function (15.35). It consists of a large number of terms.

If one assumes that τ is much larger than the width of the autocorrelation function $R_n(\delta\tau)$, one can simplify the calculation of the noise reduction factor in (15.35). The assumption is valid if, for example, $\tau \geq 50\,\text{ns}$ and at the same time the bandwidth of the noise is in the order of 100 MHz. This is applicable in a system with length of at least 10 m. In this case, in (15.35) all terms involving τ can be neglected since $R_n(\delta\tau)$ is very small for large τ and one can get

$$R_{N_3}(\delta\tau) \approx 2(\psi P_0)^2 R_{dd}(\delta\tau) R_-(\delta\tau) \exp\{-2a^2[R_n(0) - R_n(\delta\tau)\cos(\omega_0\delta\tau)]\}$$

(15.36)

A minimum value for the noise reduction factor was calculated by summing the power in the baseband contributed by the modulating noise (this is equivalent to the assumption of having a single DC-component, as in the case of single tone high-frequency modulation). Two different power spectral densities for the modulating noise were assumed: flat (band-limited) and Lorentzian, both with the same equivalent bandwidth. On Fig. 15.7 the noise reduction factor vs. τ is shown. It can be observed that the NRF decreases as the bandwidth of the modulating noise is increased. For large values of τ, the NRF becomes independent of the bandwidth.

The noise reduction factor vs. a is shown on Fig. 15.8. Note that the spectral shape of the band-pass noise is insignificant for the overall interferometric noise reduction. Thus, for large τ, the noise reduction depends only on the total power of the modulating band-pass noise.

Thus, by picking a noise generator for driving the phase modulator, one can be assured of the elimination of the interferometric noises originating from any multiple reflections. With a noise bandwidth of several hundred megahertz, centered for example at 1 GHz, interferometric noises resulting from reflections from interfaces longer than 1 m can be practically eliminated (Fig. 15.7).

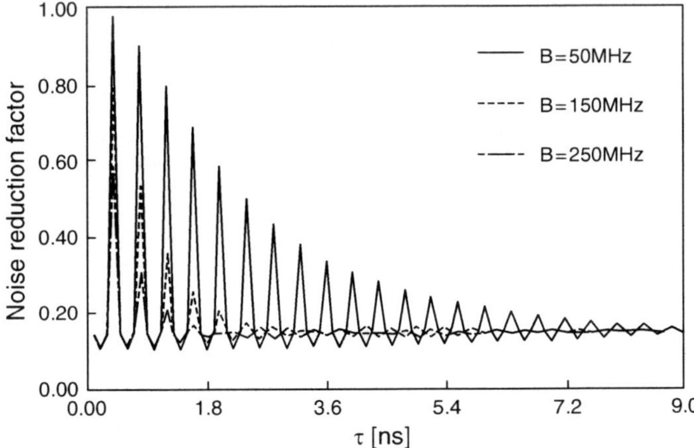

Fig. 15.7. Noise reduction factor vs. τ for external phase modulation with Gaussian noise of various bandwidth B. The phase modulation index is $a^2 = 4$ and $f_0 = 1$ GHz

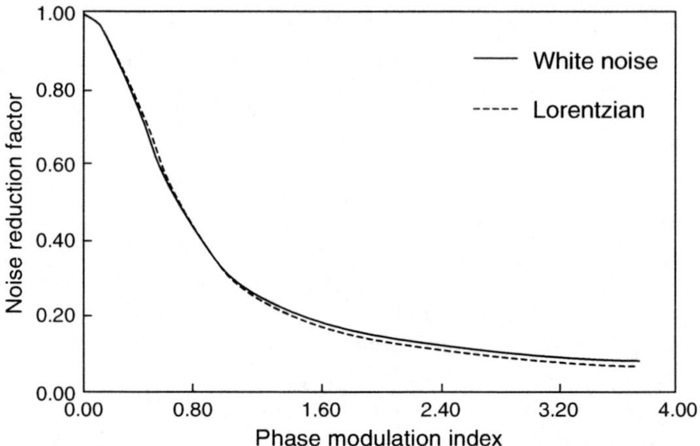

Fig. 15.8. Noise reduction factor with Gaussian noise superimposed modulation for two different power spectral densities (flat and Lorentzian), vs. a, for large τ and $R_n(0) = 1$

15.3 Multimode Fiber: Modal Noise

In the previous sections the noise reduction characteristics for the case of double reflection in a singlemode fiber was derived. The result can be extended to multimode fibers which exhibit modal noise. Extensive analysis of the modal noise phenomenon and the steps to take to prevent it was done in [156–158].

The modal noise is the result of two causes: first, any mechanical distortion of the fiber due to vibrations, bending, etc., will produce phase changes between fiber modes. Second, any source wavelength change will produce changes in the relative mode delays. One should not consider long term changes due to wavelength drift of the laser, but instead consider short term wavelength fluctuations, arising from the finite linewidth of the laser. Here one can assume that the modal noise is caused exclusively by the laser wavelength fluctuations, although the extension to both wavelength fluctuations and mechanical distortions can be easily incorporated.

Let the electric field at distance z be

$$E_{\text{out}}(t) = \sum_{i=1}^{M} \psi_i E(t - t_i) \tag{15.37}$$

where t_i are the delay times for fiber modes. To simplify the analysis, one can assume that all fiber modes are equally (uniformly) excited, which represents, in the case of no superimposed modulation, the worst case as far as modal noise is concerned [157]. This approximation does not affect the results significantly. The exact excitation will produce the same results as with uniform excitation, but with smaller number of modes. Since the number of modes in a multimode fiber is usually very large, the effect of this approximation is negligible.

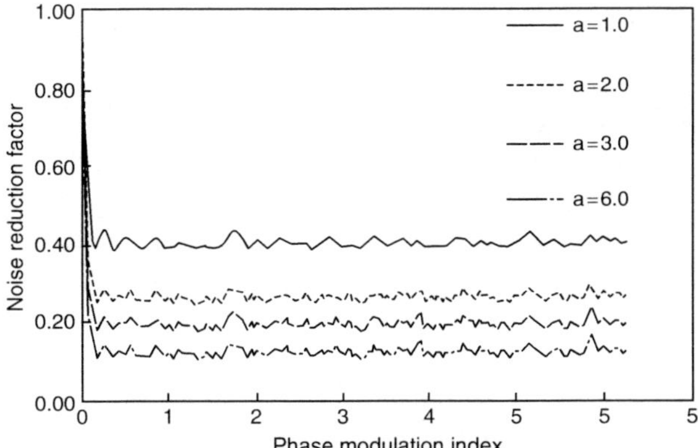

Fig. 15.9. Noise reduction factor vs. $f_0 z$ (z is the fiber length) for direct phase modulation with single tone and multimode fiber exhibiting modal noise. The number of modes is assumed to be 50 which are equally excited. The large number of modes makes the NRF aperiodic, as in the case of modulation with band-pass Gaussian noise

A high frequency superimposed modulation is applied directly to the laser as illustrated in Fig. 15.3c. Then, the autocorrelation function becomes (for direct modulation of LD with a single laser mode):

$$R_{N_0} = \frac{2P_0^2}{I_0^2(a)} R_-(\delta\tau) \sum_{i=1}^{M}{}' \sum_{j=1}^{M}{}' (\psi_i \psi_j)^2$$
$$\times \left[I_0^2 \left(a \cos \left(\frac{\omega_0 \tau_{ij}}{2} \right) \right) J_0^2 \left(2a \sin \left(\frac{\omega_0 \tau_{ij}}{2} \right) \right) \right.$$
$$\left. + 2I_1^2 \left(a \cos \left(\frac{\omega_0 \tau_{ij}}{2} \right) \right) J_1^2 \left(2a \sin \left(\frac{\omega_0 \tau_{ij}}{2} \right) \right) \right] \qquad (15.38)$$

where $\tau_{ij} = t_i - t_j$ is the relative mode delay between fiber modes and the prime denotes that the summation excludes the cases when $i = j$.

Figure 15.9 shows simulation results using (15.38) for the NRF due to directly superimposed modulation. One can notice that the increased number of fiber modes removed the periodicity that was observed in the case of double reflections in a singlemode fiber (Fig. 15.5). After the initial decay, NRF remains almost constant and independent of the frequency of the superimposed modulation. The reduction is dependent on a parameter "a", which is a modulation depth parameter.

15.4 Conclusion

The analysis of this chapter provides a theoretical framework for the suppression of interferometric noise by a superimposed high frequency modulation. The fundamental mechanism for this reduction process is the redistribution of noise energy to high frequencies due to a superimposed phase modulation, as illustrated by the simple, but idealistic case of Fig. 15.3b. The more practical situation in which the high frequency modulation is applied directly to the laser diode modulation can be interpreted as a manifestation of the ideal phase modulation scheme through frequency chirping of the lasing emission. Modulation with a single tone produces noise suppression for most situations except when the modulation frequency and the inverse round-trip delay of the reflective interfaces are related by integer multiples. This situation is avoided if modulation is applied with multiple tones, or preferably a bandpass filtered white Gaussian noise source, as long as the bandwidth of the noise source is larger than the inverse of the shortest τ anticipated in the fiber link.

In the above analysis the fiber chromatic dispersion due to the chirping has not been considered. It will manifest itself as additional power penalty, however the noise floor due to the multiple reflections observed in [69] will be avoided. In any event, the chromatic dispersion is not a dominant effect in multimode or short fiber links.

It is interesting to note that the high frequency modulation occurs naturally for a self-pulsating laser diode [158]. Although the self-pulsating mechanism originates from an undamped relaxation oscillation process due to the presence of a saturable absorber in the laser (one but not the only means to generate self-pulsation), and is quite different from the case of an externally applied modulation, the characteristics of the laser emission is practically very similar for both cases. Self-pulsation can be regarded, from the point of view of its output characteristic, as a directly modulated non-self-pulsating laser with a near 100% modulation depth. The results obtained in this paper are thus applicable to those lasers as well.

On the contrary to single-mode fibers, for multimode fiber, the NRF does not show the periodic peaking even for a single tone phase modulation. This is attributed to the large number of modes, which introduces yet another degree of randomness with consequences similar to the case of multitone modulation.

In the simulation uniform excitation of fiber modes was assumed, which gives, according to the literature, the largest modal noise in the absence of superimposed modulation. When one applies the high frequency superimposed modulation, the uniform excitation becomes the best case, for the reasons explained earlier. If combined with other existing techniques, superimposed modulation can considerably reduce the level of phase (and modal) noise in optical communication systems using multimode fiber. These techniques are useful to the extent that most legacy in-building fiber infrastructure are of the vintage multimode type, subjected to performance degradations from modal noise effects.

mm-Wave Signal Transport over Optical Fiber Links by "Feed-forward Modulation"

16.1 Principle of "Feed-forward Modulation" for mm-Wave Signal Transport over an Optical Carrier

Development of optical modulation techniques at frequencies of tens of gigahertz is motivated to a large part by potential applications in microwave and millimeter-wave signal transport in optical fiber, such as point-to-point free space data links for commercial applications and phased array antenna systems for military applications. Most of these systems operate over a relatively narrow band ($\sim$a few gigahertz), although the carrier frequency may vary over a wide range (as in frequency hopping schemes). Recent work in high-frequency, narrow-band modulation include direct modulation, specifically resonant modulation (Part II of this book), which takes advantage of coupling between longitudinal modes of a laser diode to enhance the optical modulation efficiency of laser diodes over a narrow band around the frequency separation between longitudinal modes of the laser diode, in the 10's to even 100's GHz range for common laser diode structures [88, 159]. Another possible approach is traveling wave external modulators (Appendix C). While the former has the potential of driving down component cost by utilizing existing mass manufacturing infrastructure serving the telecom industry. An inherent drawback is that, once implemented, the monolithic device does not allow for flexibility in varying the operational frequency band. This chapter describes a scheme for narrow-band ($\sim$gigahertz's) millimeter-wave optical modulation where the modulation frequency band can be *easily* tuned over a span of tens of gigahertz. NO high frequency optoelectronic components are required, with the exception of a high-speed photodetector (which is a commercially available component, and which is needed at the receiving end of *any* fiber link for transmission of mm-wave subcarrier signals regardless of how the transmitter is implemented). The scheme is based on the (electrically) tunable beat note generated by photomixing of two single frequency (DFB) laser diodes. Attempts have been made in the past to use the photomixed beat note as a variable frequency RF signal source, but the desirability of employing rugged and inexpensive laser diodes, or even the less rugged and more expensive diode-pumped YAG lasers for this purpose invariably results in instability of the beat note due to the intrinsic phase noise of the lasers, as well as extrinsic factors, remains to be an issue. In any event, the beat note so generated is only a relatively noisy RF-carrier (modulated on an optical beam) that must be encoded with the signal one intends to transmit. An well known method to achieve this is through the use of a phase-locked loop, although it is inherently difficult to extend the bandwidth of any feedback technique to much beyond 100 MHz [160]. The technique described in this chapter starts with a standard (noisy) laser beat note generated by photomixing two commercial DFB laser diodes, and, through the use of feed-forward compensation, imprints the desired mm-wave signal on the optical beam.

The operational principle of the modulation technique is shown in Fig. 16.1. The polarization controller allows for proper alignment of the polarization of the two lasers such that their electric fields are combined in the fiber optic directional coupler. The electric field of laser 1 and 2 are expressed as

$$E_{1,2} = A_{1,2}e^{i[\omega_{1,2}t+\phi_{1,2}(t)]} \tag{16.1}$$

where A is the field amplitude, ω the optical frequency, and ϕ the time dependent phase fluctuations that arises from the finite linewidth of the lasers, then the intensity propagating on the fibers after the coupler is

$$I = I_0[1 + k\cos(\omega_b t + \Theta(t))] \tag{16.2}$$

where $k = 2A_1A_2/A_1^2 + A_2^2$ is the modulation depth of the beat note, $\omega_b = |\omega_1 - \omega_2|$ is the beat note frequency, and Θ is the beat note phase. One of

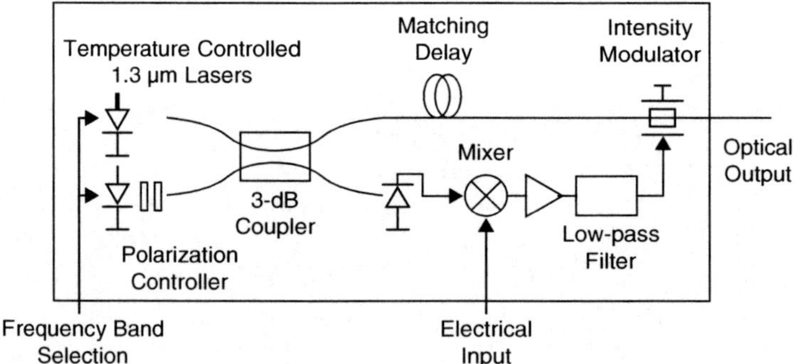

Fig. 16.1. Schematic drawing showing the operational principle of the modulation technique. The light from the two lasers is combined in the fiber optic coupler. One output of the coupler is detected and mixed with the electrical input. The down-converted signal drives the optical modulator to produce an intensity spectrum that contains a replica of the input signal that is not influenced by the beat note phase noise

the outputs from the coupler is detected by a standard high-speed photodetector and the beat note is mixed with the input millimeter-wave signal to be transmitted, which, for the purpose of this analysis, is assumed to be a sine wave at the frequency ω_{in}. The laser frequencies are adjusted so that the ω_b, is close to, but not exactly equal to ω_{in}. The down converted signal is filtered, amplified and sent to drive an optical modulator, which modulates the light from the other coupler output. The modulator only has to respond to the low frequency down converted signal, not the high-frequency mm-wave signals. Assuming the modulator is operated in the linear region, its output can be expressed as

$$
\begin{aligned}
I_{out} = \frac{I_0}{2}\{&1 + k\cos[\omega_b t + \Theta(t)] - m\cos[(\omega_{in} - \omega_b)t + \Theta(t + \tau)] \\
&- \frac{km}{2}(\cos[(2\omega_b - \omega_{in})t + \Theta(t + \tau) + \Theta(t)]) \\
&+ \cos[\omega_{in}t + \Theta(t) - \Theta(t + \tau)])\}
\end{aligned}
\tag{16.3}
$$

where I_0, is the optical intensity at the modulator input, m is the RF modulation depth of the optical modulator, τ is the difference in time delay of the two paths from the coupler to the modulator, and ω_{in} is the angular frequency of the input signal. The intensity spectrum of (16.3) has four components: the low frequency down converted signal $(\omega_b - \omega_{in})$, the beat note (ω_b), a mixer product at $2\omega_b - \omega_{in}$ and a replica of the input signal (ω_{in}). This last component at ω_{in} is the desired signal to transmit. The other spectral components will be filtered out electrically at the receiver end.

The spectral components in (16.3), are, with the exception of the $\omega_b - \omega_{in}$ component, illustrated in Fig. 16.2a. The noise pedestal on the signal at win results from a mismatch in time delay between the two arms of the feed-forward modulator. To quantify the influence of the delay, the power spectrum of the win component of (16.3) is calculated. Assuming that the laser phase noise has a Gaussian distribution with a coherence time t_b (i.e., the beat note is a Lorentzian with a FWHM linewidth of $1/\pi t_b$, which is approximately equal to twice the optical linewidth of the lasers), the power spectrum is

$$
S(\omega) = 2\pi e^{\frac{|\tau|}{t_b}}\delta(\omega) + \frac{2t_b}{\omega^2 t_b^2 + 1} \times \left\{1 - \left(\frac{|\tau|}{t_b}\frac{\sin(\omega|\tau|)}{\omega|\tau|} + \cos(\omega|\tau|)\right)e^{\frac{|\tau|}{t_b}}\right\}
\tag{16.4}
$$

where ω is the frequency offset from ω_{in}. As expected, the expression shows that the spectrum is composed of the signal (the delta function) plus a noise pedestal that will be reduced to zero when $\tau = 0$ as shown in Fig. 16.2b. Assuming $\tau \ll t_b$, the spectrum in the vicinity of the signal can be simplified to

$$
S(\omega) = 2\pi e^{\frac{|\tau|}{t_b}}\delta(\omega) + \frac{|\tau|}{t_b}
\tag{16.5}
$$

The signal-to-noise ratio in a 1 Hz bandwidth (S/N) is then simply given by $2\pi t_b/|\tau|^2$, which evaluates to 166 dB with $t_b = 160\,\text{ns}$ (corresponding to an

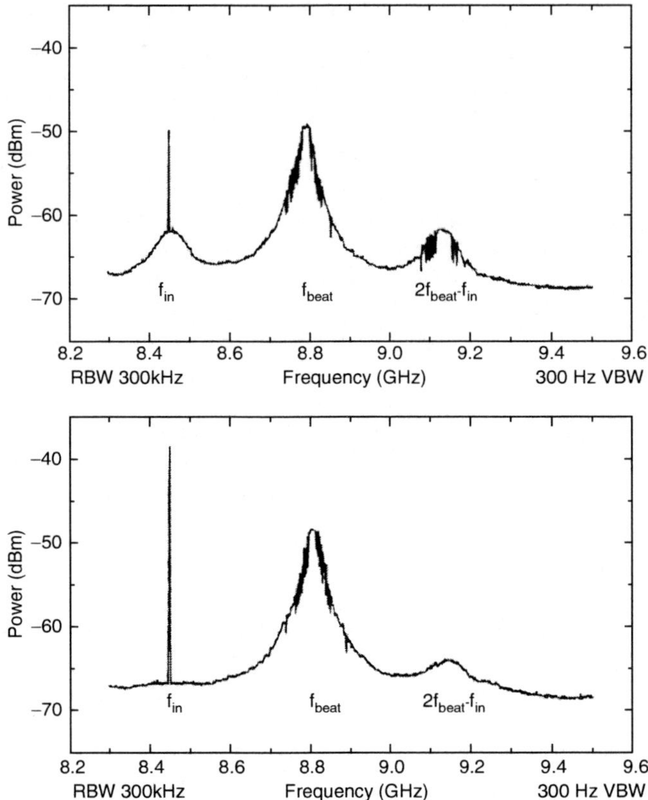

Fig. 16.2. Typical intensity spectra of the output signal showing the effect of varying the time delay difference (τ) of the two paths between the directional coupler and the optical modulator. The time delay difference, which is roughly 20 ns in (top) is reduced to less than 1 ns in (bottom), leading to the elimination of the noise pedestal, a higher signal power and a more dispersed component at $2\omega_b - \omega in$. Note the horizontal frequency range can *easily* be shifted by over 10–100s of GHz by electrical tuning of the emission wavelength (frequency) of the laser diodes, subjected only to the availability of mm-wave mixer components, and photodiodes responsive at those frequencies

optical linewidth of 1 MHz which has been demonstrated for DFB semiconductor lasers [161]) and $\tau = 5$ ps (corresponding to roughly 1 mm of propagation length in a silica waveguide). It follows that in a well-designed system with proper adjustment of τ, the noise pedestal should become insignificant compared to typical noise levels in fiber optic transmission systems.

One observes from Fig. 16.2 that, in addition to the noise pedestal, another source of noise arises from the Lorentzian laser beat note at ω_b, the fringe of which extends out to the signal frequency ω_{in}. The *S/N* due to this noise source is approximately

$$S/N = \frac{m^2(\omega_{\mathrm{in}} - \omega_b)^2 t_b}{8B} \tag{16.6}$$

With $m = 0.9$, $\frac{\omega_{\mathrm{in}} - \omega_b}{2\pi} = 20\,\mathrm{GHz}$ and $t_b = 160\,\mathrm{ns}$ as before, this evaluates to $144\,\mathrm{dB}$ $(1\,\mathrm{Hz})$, again comparable to those of traditional fiber optic transmission systems. This also illustrates the importance of using narrow line width lasers for this transmission scheme.

The above noise components present in this modulation scheme are in addition to shot noise and thermal noise (laser relative intensity noise is not important because the frequency range of operation is well above the laser relaxation oscillation frequency). These latter noise sources are common to all fiber optic links and will not be discussed further here.

For an experimental demonstration two $1.3\,\mu\mathrm{m}$ DFB lasers were used with thermoelectric coolers for temperature control and stabilization. Using a combination of temperature and injection current tuning, the beat note could *easily* be tuned over the $45\,\mathrm{GHz}$ range of the detection bandwidth of the photo detectors. Under typical bias conditions, the beat note intensity spectrum of the two lasers can be accurately modeled as a Lorentzian with a full-width-at-half-maximum (FWHM) of $25\,\mathrm{MHz}$ (corresponding to a coherence time of $13\,\mathrm{ns}$) near line center, but had roughly $5\,\mathrm{dB}$ of extra noise more than $2\,\mathrm{GHz}$ away from the peak. The optical modulator had a bandwidth of $4\,\mathrm{GHz}$, and a variable delay was used to reduce the delay difference of the two optical paths to less than $0.25\,\mathrm{ns}$.

With these components, an S/N of $103.3\,\mathrm{dB}$ $(1\,\mathrm{Hz})$ at $\frac{\omega_{\mathrm{in}} - \omega_b}{2\pi} = 8\,\mathrm{GHz}$ was achieved. The ratio of the energy in the sine wave to the energy in the beat note is roughly 0.2, which implies that the modulation depth is approximately 0.9. Considering the $5\,\mathrm{dB}$ extra noise of the laser beat note compared to a Lorentzian, this result is in reasonable agreement with (16.6), which gives an S/N of $113\,\mathrm{dB}$ with $m = 0.9$, $\frac{\omega_{\mathrm{in}} - \omega_b}{2\pi} = 2\,\mathrm{GHz}$ and $t_b = 13\,\mathrm{ns}$.

A rudimentary demonstration of the high frequency capability of this modulation technique was carried out through transmission of a simulated radar pulse train at a carrier frequency of $40\,\mathrm{GHz}$, as shown in Fig. 16.3. The repetition rate of the radar pulse is $247\,\mathrm{kHz}$ and the duty cycle is approximately 25%. The S/N at $40\,\mathrm{GHz}$ is thermal noise limited due to the large noise figure of the mixer used to down convert the output signal. This leads to some degradation of the transmitted signal, but the comparison in Fig. 16.3 clearly shows that the output replicates the input closely. With proper choice of components, the S/N at these millimeter-wave frequencies should be close to that shown before [$>140\,\mathrm{dB}$ $(1\,\mathrm{Hz})$].

In principle, the modulation technique described above is capable of a transmission band centered at the beat note frequency with a width equal to twice the bandwidth of the optical modulator. In this band the response is determined only by the optical modulator, the receiver and the electrical components. This is illustrated in Fig. 16.4 which shows the one-sided normalized frequency response with the beat frequency set to $36\,\mathrm{GHz}$. The response has

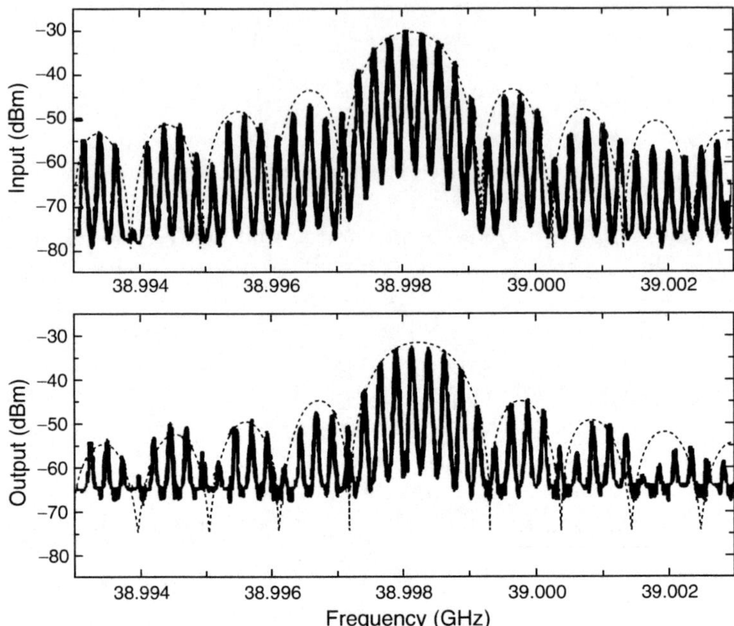

Fig. 16.3. Input and output spectra of a simulated radar pulse at 40 GHz, showing transmission of good fidelity

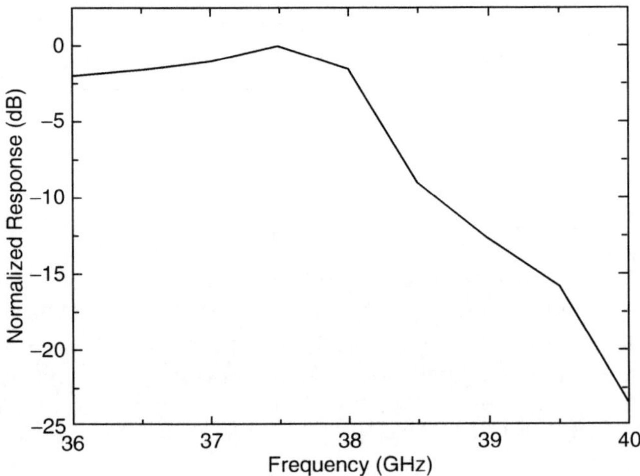

Fig. 16.4. Normalized frequency response with the beat note at 36 GHz. The sharp roll-off is due to the IF-bandwidth of the down converting mixer at the output and the preamplifier of the optical modulator

a relatively flat region from 36 to 38 GHz, followed by a sharp roll-off caused by a combination of the IF-bandwidth of the down converting mixer at the output and the bandwidth of the amplifier driving the optical modulator. In a practical system, however, the center of this band is not useful, because of the noisy beat note, but bandwidths of several gigahertz are still possible.

This section describes an innovative technique for band limited, tunable, high frequency optical modulation in the mm-wave frequency range. Experimentally, optical transmission at 40 GHz has been demonstrated with good fidelity for a useful bandwidth in excess of 1 GHz and an S/N of 103.3 dB (1 Hz) at 8 GHz away from the mm-wave carrier using standard 1.3 μm DFB lasers and a 4 GHz external optical modulator. All components used are off-the-shelf, and integration of the components to perform the intended function is through rugged fibers and fiber couplers is a well known art. The operational mm-wave band is tuned electronically, with no adjustable moving parts. Theoretically, it can be shown that with very stable DFB lasers ($\lesssim$1 MHz line width) and a high frequency (20 GHz) external optical modulator (which is commercially available), the signal-to-noise ratio can be improved to better than 140 dB (1 Hz). Further improvements can be achieved by using stable, high power lasers like diode pumped YAG, albeit at the cost of increased system cost and complexity. Because of the capability for high frequency operation, wide tunability range and good signal quality, this modulation technique is a viable alternative to common direct or external modulation techniques in remote-antenna communication and/or radar systems operating at the mm-wave band.

16.2 Demonstration of "Feed-Forward Modulation" for Optical Transmission of Digitally Modulated mm-Wave Subcarrier

The use of optical fiber for efficient transport of narrowband mm-wave signals is a subject of considerable interest for a variety of applications, both commercial and military [107, 110, 162–164]. Recently, direct modulation up to 30 GHz has been demonstrated for quantum-well lasers [165]. The principles and physics which govern these limits have been discussed thoroughly in Part I of this book. In Part II, narrowband resonant enhancement has been described and demonstrated for signal transmission at frequencies $\geq$35 GHz [162, 163]. A new technique was introduced for mm-wave signal transmission using feed-forward optical modulation. Its basic small-signal modulation and noise characteristics was described in Sect. 16.1 [164]. The potential of this technique to transmit data with high fidelity is conditional upon having adequate small-signal bandwidth as well as good distortion characteristics under large signal modulation. In this section, a demonstration of transmission of 300 Mbit s^{-1} BPSK data at a carrier frequency of 39 GHz over 2.2 km of singlemode fiber is described. Furthermore, these results imply that gigabit-per-second data

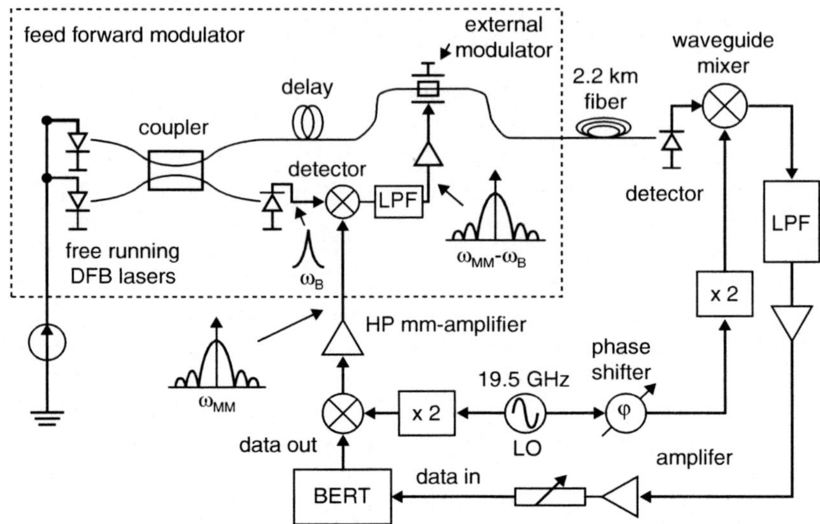

Fig. 16.5. Experimental setup used to measure BER

rates at subcarrier frequencies extending to 100 GHz and beyond are possible using currently available commercial components.

The experimental setup is shown in Fig. 16.5. The dotted enclosure represents the feed-forward mm-wave optical modulator. Details on the operational principle of this modulation technique was described before in Sect. 16.1 [164]. In brief, a beat note at 36.5 GHz generated by photomixing two 1.3 μm DFB lasers is electrically mixed with the input mm-wave signal which centers at 39 GHz and is binary-phaseshift-keyed (BPSK) modulated at 15–300 Mbit s^{-1}. The resulting IF "error" signal at $39.0 - 36.5 = 2.5$ GHz, is fed-forward to an external optical modulator, which impresses the BPSK modulated mm-wave carrier at its output. The optical signal out of the feed-forward modulator is transmitted through fiber and detected by a high speed photodiode. Bit-error-rate (BER) measurements are carried out after the signal is downconverted to baseband, amplified, lowpass filtered and Sent to the error-rate tester. Figure 16.6 shows the measured BER as a function of received optical power for transmission distances of several meters (squares) and 2.2 km (crosses) at 150 Mbit s^{-1}. The BER against CNR is also shown in Fig. 16.6 (circles). The CNR was obtained by measuring the baseband signal-to-noise ratio (SNR) and using CNR = 2SNR for BPSK [166]. At BER = 10^{-9} an optical power of -9.8 dBm and a CNR of -15 dB is required. To demonstrate transmission at >150 Mbit s^{-1}, one can plot the optical power penalty as a function of bit rate, with BER fixed at 10^{-9}. As shown in Fig. 16.7, an additional 1.2 dBm of optical power is required to transmit at 300 Mbit s^{-1}.

As discussed in Sect. 16.1, two noise components specific to the feed-forward transmitter are the delay mismatch between the two paths from the

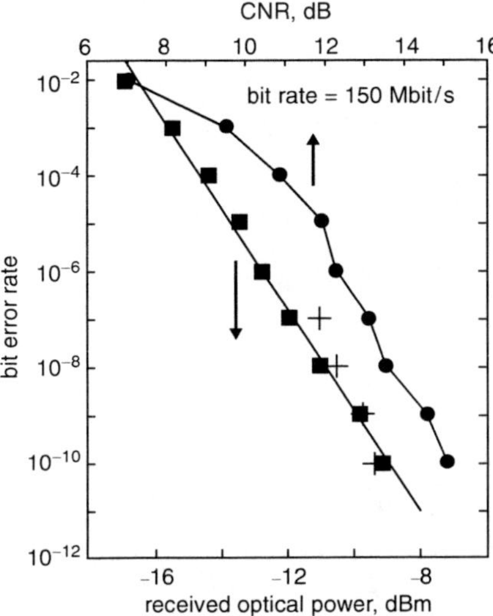

Fig. 16.6. BER as a function of received optical power and CNR at $150\,\mathrm{Mbit\,s^{-1}}$. CNR measured to within $\pm 1.5\,\mathrm{dB}$ accuracy

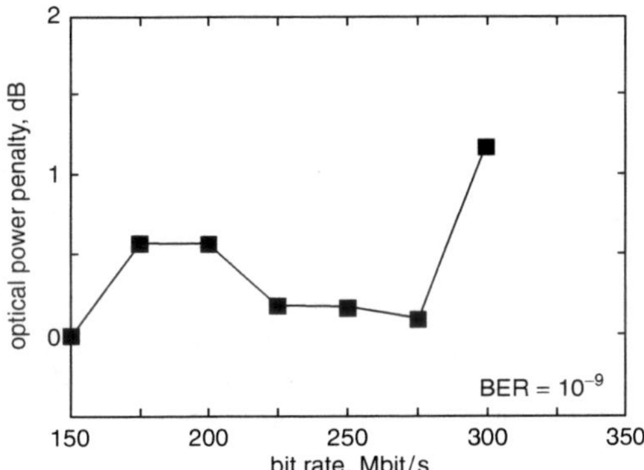

Fig. 16.7. Optical power penalty as function of bit rate

coupler to the external modulator (which creates a phase noise pedestal at ω_{MM}), and the "spill-over" of the "tail" of the beat note (at ω_B) into the signal band at ω_{MM} [164]. The relative importance of these noises, as compared to conventional shot and thermal noise, can be quantified using the following expression for the CNR:

$$C/N = [(C/N)_{\mathrm{shot}}^{-1} + (C/N)_{\mathrm{thermal}}^{-1} + (C/N)_{\mathrm{delay\ mismatch}}^{-1} + (C/N)_{\mathrm{beat\ note}}^{-1}]^{-1}$$

$$\times \left(\frac{16qB}{RP_0 k^2 m^2} + \frac{32k_B T B F}{R^2 P_0^2 k^2 m^2 R_0} + \Delta v |\tau|^2 B + \frac{2\Delta v B}{\pi m^2 f_D^2} \right)^{-1} \tag{16.7}$$

Table 16.1. Summary of measured system parameters

Parameter	Symbol	Value
Received optical power	P_0	$-9.8\,$dBm for $BER = 10^{-9}$
Modulation depth of beat note	k	~ 1
Modulation index of modulator	m	~ 1
Detector responsivity	R	$0.4\,$A/W
Resistance	R_0	$50\,\Omega$
Filter bandwidth	B	$300\,$MHz
Noise figure	F	13
Beat-note linewidth	Δv	$30\,$MHz
Delay mismatch	r	$<5\,$ps
Difference frequency	$f_D = f_{MM} - f_B$	$2.5\,$GHz
Carrier power	C	$R^2 P_0^2 k^2 m^2 / 32$
C/N_{shot}	$RP_0 k^2 m^2 / 16qB$	$47\,$dB
C/N_{thermal}	$R^2 P_0^2 k^2 m^2 R_0 / 32 k_B TBF$	$22\,$dB
$C/N_{\text{delay mismatch}}$	$1/\Delta v |\tau|^2 B$	$66\,$dB
$C/N_{\text{beat note}}$	$\pi m^2 f_D^2 / 2\Delta vB$	$30\,$dB

The laser relative-intensity-noise (RIN) is not included here because the operating frequency range is well above the relaxation frequency of the lasers. The parameters in the above expression are defined and summarized in Table 16.1. Note that the existing system is thermal-noise limited because the received optical power is relatively low. The error probability for BPSK is therefore $\text{BER}_{\text{BPSK}} \approx 1/2\,\text{erfc}[(C/N)_{\text{thermal}}]$.

From the above discussion, it is clear that the photomixed beat note can be easily adjusted to any frequency within a few hundred gigahertz. The maximum carrier frequency in the feed-forward transmitter is therefore limited only by the bandwidth of the photodetector ($40\,$GHz in our case). The data rate on the other hand is limited by the electrical bandwidth of the external modulator ($4\,$GHz in our case). Photodetectors with bandwidths in excess of $100\,$GHz have been demonstrated and external modulators with bandwidths $\geq 20\,$GHz are commercially available. Therefore, gigabit data rates at carrier frequencies $>100\,$GHz are possible. As an example, for transmission of a $5\,\text{Gbit s}^{-1}$ signal centered at $f_{MM} = 100\,$GHz one would use an external modulator with a bandwidth of $10\,$GHz. The beat note should then be adjusted such that $f_B - f_{MM}$ is $5\,$GHz (to allow for lower and upper sidebands of the data signal).

Optical transmission of $300\,\text{Mbit s}^{-1}$ BPSK data at $39\,$GHz using feed-forward modulation have been demonstrated. The modulator is based on an innovative principle which allows for gigabit data transmission at carrier frequencies well beyond the relaxation frequency of a semiconductor laser. The modulator can be built with commercially available components that are all optoelectronic-integrated circuit (OEIC) compatible.

Part IV

Appendices

A

Notes on RF Link Metrics

A.1 Notes on Relations Between Distortion Products, Noise, Spur (Spurious) Free Dynamic Range (SFDR)

SFDR is a common parameter used to characterize the fidelity of an analog (microwave) device or link. It is expressed in the rather unusual unit of $dBHz^{-2/3}$. It has been encountered before in Fig. 10.4, a more detailed explanation is given here.

Figure A.1 shows a general plot of the RF output (power, in dBm) versus RF input power into a "device", the "device" can be a single or a combination of component such as RF amplifiers mixers and attenuators, or a fiber optic link, where the RF input is measured at the input to the optical transmitter and the RF output at the output of the optical receiver. For the sake of simplicity assume two primary RF tones at frequencies ω_1 and ω_2 are present at the input to the "device". At the output of the "device" one nominally observes the two primary RF tones at frequencies ω_1 and ω_2. These are designated as "signals" since they are the ones present at the input of the "device", and which one wishes to convey to the output of the same "device", verbatim. The ideal situation is that the recovered RF at the output of the "device" tracks the input RF. This is represented by the straight line denoted by "signal" in the plot, the linearity of the "signal" line merely means that the input signals are not "compressed" even at high input RF power levels; "compression" represents a very severe form of distortion of the signals; most RF circuit and system functions require distortions to be avoided at a much lower level. These are represented by the second and third order harmonic as well as intermodulation distortions, as discussed in Chap. 3 previously. The power level of second harmonic distortions increases as the square of the input power, as it originates from a product of two primary tones, the slope of the second order distortion terms is 2. By the same token, the third order distortion terms have a slope of 3. Also shown in Fig. A.1 is the noise background level at the output of the device, originating from either the input to the device or the device

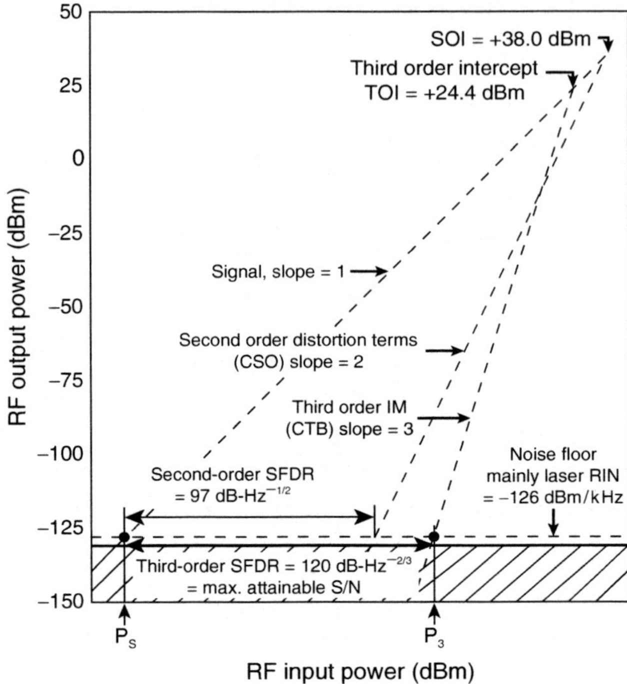

Fig. A.1. An illustration of fundamental, second order, and third order intermodulation products showing Spurious-Free Dynamic Range (SFDR)

itself. All distortion terms are undesirable, as well as noise. Second order distortions such as "CSO" as described in Appendix A.2 originate from existence of a term in the transfer function of the "device", which is the square of the input. Likewise, third order distortions such as "CTB" arises from terms of the third power. As a result, at the output of the "device", the input-output relationships of "CSO" (second order) and "CTB" (third order) components have slopes of 2 and 3 respectively.

As the input signal power increases, the S/N ratio increases proportionally (linearly on a dB scale); while the CSO and CTB (representing second and third order distortions) increases with slopes of 2 and 3, respectively. The point where the "CTB" line intersects the "signal" line is commonly known as third order intercept (TOI) point. At this point the S/N ratio at the output is 1, if one regards distortion as a form of noise. At the input signal power level where the third order distortion penetrates above the noise background a maximum S/N ratio is achieved (see Fig. A.1). Further increase in input signal power does not bring forth an improvement in S/N ratio, if one regards distortions as a form of noise. A simple geometric examination of Fig. A.1 reveals that the input range where a signal can be detected without detecting CTB (CSO) distortions – i.e., the difference between input power levels where the output signal penetrates above the background noise (P_S in Fig. A.1), and where the second (third) order distortions penetrate above the background noise (P_2, P_3 respectively in Fig. A.1). The quantity $P_3 - P_S$ is also known as the "Spur(spurious)-Free Dynamic Range" (SFDR – "3rd. order SFDR", to be exact.) The Second Order SFDR is defined similarly. Simple geometry also lead to the conclusion that the quantity SFDR represents the maximum S/N that can be achieved regardless of input RF power.

Every "dB" increase in detection bandwidth of the system results in a corresponding "dB" increase in background noise floor – if one assumes the noise spectrum is flat at the frequency range of operation. Every "dB" increase in detection bandwidth of the system thus results in a corresponding *increase* in P_S and an increase in P_3 of $1/3\,$dB (since the CTB line has a slope of 3), resulting in *a net change in* SFDR $= P_3 - P_S$ of $-2/3\,$dB. Therefore the SFDR $= P_3 - P_S$ varies as the $-2/3$ power of detection bandwidth of the system. The unit of SFDR is therefore expressed as dB-Hz$^{-2/3}$.

Using a similar reasoning as above, the second order SFDR is expressed as dB-Hz$^{-1/2}$.

A.2 Notes on Intermodulation Distortion in a Multi-Channel Subcarrier Transmission System: CTB and CSO

A.2.1 Composite Triple Beat (CTB)

Chapter 3 and Sect. A.1 discussed distortions generated at the output of the intensity modulated laser due to two closely spaced RF modulation tones, say, f_1 and f_2, in the form of "spurious" images at both sides of the two primary tones, namely at $2f_1 - f_2$ and $2f_2 - f_1$. These are third order distortion products due to a third order non-linearity in one of the elements comprising the link, of which the laser contributes substantially. These are called third order intermodulation (IM) distortion products.

When more than two, say, N primary modulating tones are present, a myriad of sum/difference beat notes result from the nonlinearity. For a system encountering third order nonlinearities, the general form of the frequencies at which distortion products are encountered is $f_A \pm f_B \pm f_C$ for *all* possible combinations of A, B and $C < N$.

For cable television systems, Composite Triple Beat (CTB) is defined as "a composite (power summation) of third order IM distortion products originating from all possible combinations of three channels which lands on a given frequency (channel). It is expressed as a ratio (in dB) of the RF carrier signal to the level of the composite of third order distortion components centered at the carrier."

In a CATV system, the performance specifications are given by the carrier-to-noise ratio (CNR), the composite-second-order (CSO), the composite-triple-beat (CTB) and the cross modulation (XM). The last three distortion

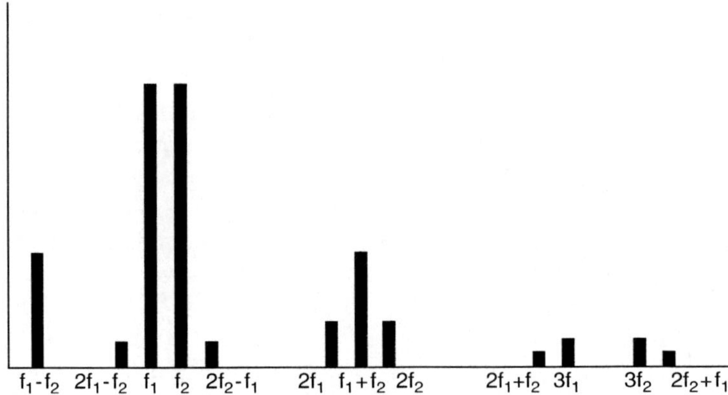

Fig. A.2. Frequency beats from a nonlinear device

terms are related to the device linearity. In the current RF CATV system, RF trunk amplifiers and external amplifiers are the primary sources of distortions. Shown in Fig. A.2 is the output spectrum of a nonlinear device output. With two input carriers at f_1 and f_2, the second-order distortions generate at frequency $2f_1$, $2f_2$, $f_1 + f_2$ and $f_1 - f_2$. The third-order distortions generate at $3f_1$, $3f_2$, $2f_1 + f_2$, etc. For a multi-carrier system, some of the second-order or third-order distortion may fall at the same frequency that is occupied by a channel. Thus the picture quality of this channel is degraded. One can define CTB as the sum of all third-order distortion powers at one channel relative to the carrier power at that channel. The CTB at f_m is then given by

$$CTB_{f_m} = 10\log\left(\frac{\sum_i \sum_j \sum_k P^{f_i \pm f_j \pm f_k}}{P^{f_m}}\right) \quad \text{where } f_i \pm f_j \pm f_k = f_m \quad \text{(A.1)}$$

Similarly, CSO at f_m can be defined as

$$CTSO_{f_m} = 10\log\left(\frac{\sum_i \sum_j P^{f_i \pm f_j}}{P^{f_m}}\right) \quad \text{where } f_i \pm f_j = f_m \quad \text{(A.2)}$$

Since the final judgment of picture quality is from the human eyes, the specifications of the CNR, CTB, CSO and XM are based on the subjective testing results. Although there is no strict regulation on the specifications, the typical current specifications in the super trunk are CNR = 52 dB, CSO = −65 dB, CTB = −65 dB and XM = −52 dB. At the subscribers end, CNR is designed to be 40–45 dB. Shown in Table A.1 is the subjective test results conducted by the Television Allocations Study Organization (TASO) [167]. Most households probably only receive a service with CNR of 35 dB. In order to accommodate the high degradation from the chained amplifiers in the distribution link, CNR value at the headend is very high. If the number of amplifiers can be reduced, such as by using low loss fiber-optic technology, the stringent requirement

Table A.1. Television Allocation Study Organization (TASO) report of subjective test results

TASO picture rating	S/N Radio
1. Excellent (no perceptible snow)	45 dB
2. Fine (snow just perceptible)	35 dB
3. Passable (snow definitely perceptible but not objectionable)	29 dB
4. Marginal (snow somewhat objectional)	25 dB

at the transmitter site may be relaxed. Due to its stringent requirements, a lightwave analog video distribution system has became a challenging area for research. Since lightwave components are quite different from the RF counterparts, further investigations are needed to optimize the system performance. Using an approach different from the present RF system design in the system architecture may be the key factor to a successful lightwave distribution system.

A full accounting of these beat notes requires a painstaking combinatorial analysis of any three frequencies among the N channels present. Interested readers may consult an Application note published by Matrix Test Equipments Inc., a company which specializes in test equipments for multi-channel testing purpose [168].

A.2.2 Composite Second Order Intermodulation (CSO) Distortion

While CTB's are a result of beating of three primary notes (two or more of which can be identical) due to a third order nonlinearity. Composite Second Order distortion (CSO) is a result of two, say f_A, f_B (can be identical) carriers experiencing a second order non-linearity. An argument similar to the third order distortion can be made here.

Suffice to say that, for a 100 channel transmission system the number of second and third order terms falling on any given channel can run into thousands. This then translates into a two-tone intermodulation requirement of -60 to 70 dB range in order for CTB and CSO requirements to be met, as laser transmitter products serving the CATV industry do [169, 170].

Based on the theoretical and experimental results of Chap. 3 it is evident that, fundamentally, distortions (of *any* type) are most severe at around the intrinsic relaxation frequency of the laser, and thus suppressing the resonance can lead to a reduction in the IM levels. Further considerations must be taken into account to allow for effects due to device imperfections, such as thermal effects, or carrier leakage from the lasing region – the former usually not a factor at frequencies above tens of MHz's. These device imperfections are mitigated most cost-effectively by using (calibrated) external nonlinear circuit elements at the input to the laser transmitter as compensators, as routinely practiced in present day commercial high performance CATV laser transmitter products . Intrinsic nonlinear distortions related to relaxation oscillation resonance are frequency dependent, as explained in Chap. 3, and thus are more difficult to mitigate by means of pre-distortion. These fundamental laser modulation distortions can only be dealt with effectively by pushing the relaxation oscillation resonance to as high a frequency as possible, away from the signal transmission bands. Thus, developments in high speed lasers like those described in Part I of this book can lead to concomitant improvements in fundamental linearity in direct modulation of laser diodes.

B

Ultra-high Frequency Photodiodes and Receivers

B.1 Ultra-high Speed PIN Photodiodes

With the recent advancement of gigabit fiber communication and the photonic distribution of microwave signals, there is a growing need for high-speed photodetectors.

Two types of photodetectors commonly used for high-speed applications are *p-i-n* and Schottky photodiodes (see Fig. B.1). In both devices, photon absorption in the depletion region of a reverse-biased junction creates electron-hole pairs. These carriers are swept out of the high-field region to create a current in an external circuit.

The speed of these photodiodes is limited by depletion region transit time and capacitance. Transit time refers to the time required for the electrons and holes to drift across the high-field depletion region. It is determined by carrier saturation velocity ($\sim 3 \times 10^6 \, \mathrm{cm \, s^{-1}}$) and the depletion region thickness t, which can vary. Depletion region thickness is thus the design parameter controlling transit time, and should be made inversely proportional to the desired bandwidth.

Capacitance slows the device via an RC time constant where the resistance is that of the device load impedance. The capacitance is proportional to the active area and inversely proportional to depletion region thickness. For high-speed operation, then, both the active area and depletion region thickness should be minimized. A small active area, however, places demanding requirements on the focusing optics for the detector, or the alignment accuracy between the fiber and the photodetector. A thin depletion region means only a fraction of the incident photons will be absorbed. To optimize speed while maintaining performance, designers generally make the active area and depletion region thickness just small enough to satisfy the speed requirements; transit time is typically made comparable to the RC time constant. Using this simple approach, engineers have designed high-speed Schottky detectors that can achieve bandwidths as high as 60 GHz.

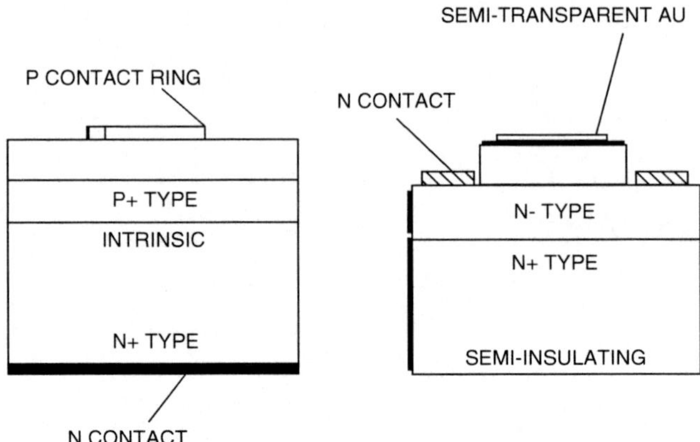

Fig. B.1. Generic illustration of photodiodes. The high-field region is the (thin) intrinsic layer in the *p-i-n* photodiode (*left*) and the *n*-region in the Schottky photodiode (*right*). These are the regions where photons are absorbed, generating electron-hole pairs, and rapidly swept to the highly conducting N^+/P^+ terminal layers. (©Bookham, Inc. Reprinted with permission)

A Schottky configuration is the faster of the two designs – in the case of a *p-i-n* diode, if the top *p*-layer is absorbing, the carriers generated in this undepleted, low-field region must diffuse out at slow speeds. Schottky photodiodes also offer lower parasitic resistance. The *n*-type Schottky diode, for example, has only an *n*-layer and no *p*-layer. In a top-illuminated *n*-type diode, the carriers are created near the top metal contact; the holes, which are the slower carriers, travel just a short distance to the metal.

The detectors have been designed for both back-side and front-side illumination. For backside illumination, light is incident through the transparent indium phosphide (InP) substrate and absorbed in the indium gallium arsenide (InGaAs) active region, permitting detection of wavelengths from 950 to 1,650 nm. The top Schottky contact serves as a mirror, allowing a double-pass through the absorbing layer to enhance quantum efficiency.

The front-illuminated devices are fabricated with both InGaAs and gallium arsenide (GaAs) absorbing layers and have a thin, semi-transparent gold (Au) Schottky metal. The high sheet resistance of the thin gold layer is detrimental to the high-speed performance, so a current-collecting ring of thick gold is added to the periphery of the active area to minimize this resistance. The devices are sensitive for wavelengths ranging from 400 to 1,650 nm.

An intrinsic photodiode designed for high-speed operation is necessary, but not sufficient, for high-speed optical detection. The bias circuitry and the high-speed connection to the 50-ohm output transmission line must also be carefully designed to produce the desired response. This response is dictated by the application and is generally either a flat frequency response, with the responsivity varying only slightly across the operating bandwidth, or a fast,

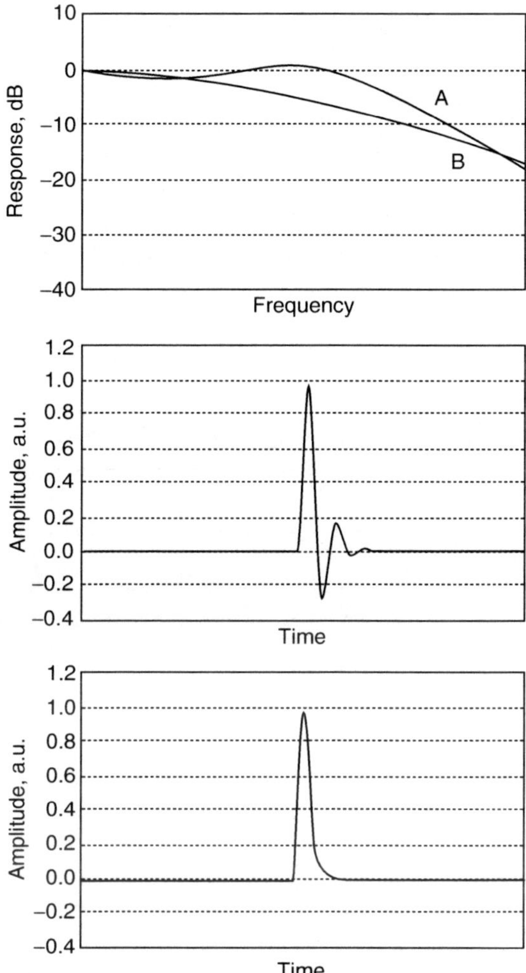

Fig. B.2. Detector designed for enhanced responsivity at high frequencies provides a nearly flat frequency response (*top, curve A*), but suffers from ringing in the temporal domain (*middle*). Detector with a clean, ring-free impulse response in the temporal domain (*bottom*) experiences roll off in the frequency domain, reducing the 3-dB frequency. (©Bookham, Inc. Reprinted with permission)

ring-free impulse response. Fourier transform techniques show that the flat frequency response suffers from controlled ringing in the temporal domain (impulse response, see Fig. B.2). The ring-free impulse response, on the other hand, corresponds to a characteristic roll-off in the frequency domain and a corresponding reduction in the 3-dB frequency.

Recently developed time-domain optimized detectors with a fast, minimal-ring-ing impulse response are especially useful for digital communications

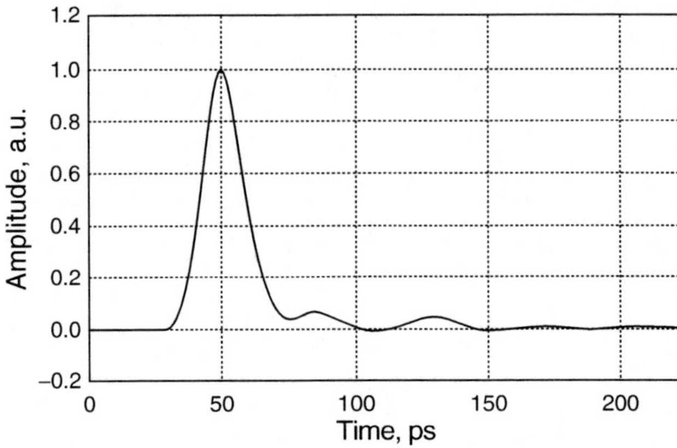

Fig. B.3. Impulse response of Bookham, Inc. Model 1,444 measured with a 50-GHz scope and a 150-fs full-width-at-half-maximum input pulse at 1.06 µm shows only slight amounts on ringing. (©Bookham, Inc. Reprinted with permission)

applications in which spurious ringing can degrade an eye diagram and bit error rate (BER). These detectors have been designed with a resistive matching network that presents the diodes with a constant 50-ohm impedance to eliminate unwanted reflections, and also terminates the detector so that its impedance is 50 ohms. The internal 50-ohm termination makes the detectors directly compatible with BER testing using switched digital hierarchy and SONET filters. The detectors are also fabricated with on-chip bias circuitry, such as integrated bypass capacitors which provide near-ideal performance to well beyond 60 GHz.

The impulse response of a detector with an 18-ps full-width-at-half-maximum shows only a slight amount of ringing (see Fig. B.3). The measurement has been made with a 50-GHz sampling oscilloscope and short (less than 200 fs) pulses from a diode-pumped neodymium-doped glass (Nd:glass) laser operating at 1.06 µm. Connecting the detector module directly to the input of the oscilloscope eliminates RF cables, and the detector's fiber-optic input then receives signals from the system under test.

Detectors with a flat frequency response can be implemented with some slight inductive peaking to enhance the responsivity at higher frequencies. Such detectors are useful for applications involving the optical transmission of microwave and millimeter wave RF signals, such as wireless cellular networks or antenna remoting in military or commercial communication satellite systems.

High-speed detectors are an important component in high-bandwidth optical communications. By optimizing the photodiode for high-speed operation, and by designing the microwave circuitry a flat frequency response can be readily achieved. These high speed photodiodes are now commercially available from companies such as Bookham, Inc [171].

B.2 Resonant Receivers

Except for very long distance fiber links where the optical signal to be detected is very weak. Most broadband optical receiver for fiber optic links today are of "pin-FET" design, a schematic of which is shown in Fig. B.4 where Z_L is typically just a (large) resistor R. The detection bandwidth of this type of receiver (typically referred to as "front-end" because its task is mainly to recover the optical input and convert it verbatim into an electrical signal to be further processed by optimal-filtering/digital electronics to follow. The bandwidth of this "front end" is determined by the RC time constant, where R is the load resistor, C is the (junction+parasitic) capacitance of the PIN photodiode, in addition to the input capacitance of the high input impedance amplifier to follow, the total equivalent C is typically very small, thus contributing to the high detection bandwidth of this "front end". Thermal noise is generated by all resistive elements, with noise power inversely proportional to the real part (resistive portion) of the impedance, the use of a high load resistor with a high input impedance amplifier such as a high frequency MESFET minimizes thermal noise contribution.

The appeal of this design is based on the very high speed and low capacitance of p-i-n diodes, described in Sect. B.1 above, together with the high load resistance which results in both a high voltage input to the gate of the FET transistor as well as a low thermal noise source from the load resistor ($\sim kT/R$). This configuration thus provides a broadband/low-noise performance for converting the input optical intensity variation into a corresponding output voltage variation.

While very useful for base-band reception, when the signal of interest spans only a limited band (as in subcarrier signal transmission) one can design an

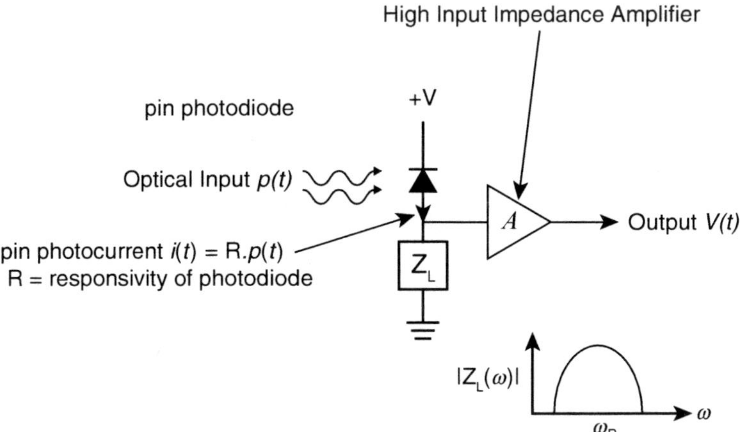

Fig. B.4. Schematic of a resonant "pin-FET" optical receiver design – the FET is represented by the high input impedance amplifier A in the figure

impedance (preferably one which does not incorporate resistive elements as such elements are sources of thermal noise) of which the magnitude $|Z_L(\omega)|$ exhibits a resonance peak at $\omega = \omega_R$ as illustrated in Fig. B.4). Such an example can be found in subcarrier resonant receiver design [172]. This well-established art in filter design can be carried over, albeit made more challenging in implementation at very high frequencies to the mm-wave band. Nevertheless, it is a subject which has been well studied within the discipline of millimeter wave devices and circuits, and will not be further addressed here.

C

High Frequency Optical Modulators

The performances and limitations of directly modulated laser diodes as optical transmitters for very high frequency (mm-wave) signals has been discussed quite thoroughly in the foregoing chapters of this book. An alternative to directly modulating the laser diode is external optical modulator for intensity modulation of the output of a continuously operating (cw) laser source. This is the preferred solution in some existing field-deployed systems where the issue of optical loss due to fiber coupling in and out of the modulator has been resolved. A prominent example being the case of a widely deployed component in telecom transmitter, the Integrated-Modulator-Laser, "IML" in which an electroabsorption modulator "EAM" is integrated with a laser source on the same chip. This device is advantageous for two reasons

1. The integrated device minimizes optical coupling loss from the laser into the modulator waveguide,
2. The issue of matching the laser wavelength to the EAM is easily resolved since both devices are fabricated on the same semiconductor wafer in one single process.

It is necessary to match the laser wavelength to the EAM device because EAM functions through varying the band-edge energy (wavelength) of the semiconductor medium by an applied voltage, thus significantly varying the amount of absorption of light passing through it *if the wavelength of the light is near the band-edge*. EAM's are produced and deployed in volume, in the form of IML's for telecom applications, where the modulation format is invariably digital on-off. EAM's has not been used extensively as an independent optical component even in digital modulation, most likely due to the difficulties in optical coupling in/out of it, in addition to the need for matching the wavelength to the laser source. Its analog modulation characteristics has not been extensively characterized either, but from existing published data the input/output (intensity absorption vs. applied voltage) characteristic is far from being linear.

The other popular type of external modulators which has been characterized extensively is of the Mach Zehnder Interferometric type (Mach Zehnder Modulator, "MZM",) fabricated in a single-chip integrated guided wave fashion on an electrooptic material such as Lithium Niobate or III-V compound [173, 174].

The operation speed of both EAM and MZM are practically limited only by electrical parasitic effects, since the intrinsic responses of the material properties responsible for modulating the optical waves (band edge energy and optical refractive index variations in response to an applied electric field for EOM and MZM respectively, can be considered instantaneous for all practical purposes. Since the principal parasitic element is the capacitance between the modulator electrodes, there are two possible means to tackle the problem.

1. Narrow-band resonant approach, by placing a parallel inductor to "tune out" the parasitic capacitance. This results in a resonance peak in the response function at the desired frequency, this approach is reminiscent of "resonant modulation" in directly modulated laser diodes, as described in Part III of this book.
2. As an alternative, much work has also been done to neutralize parasitic capacitance limitations by using "traveling wave" electrode structures, which in essence is a transmission line structure, where the parasitic capacitance (per unit length) is compensated for by the inductance (per unit length) of the electrode, resulting in a (real) characteristic impedance, high frequency signals thus do not suffer loss despite presence of the capacitive shunt.

For high speed modulation, the primary consideration for this type of "traveling wave" modulator configuration is matching of the velocity of the optical wave and that of the modulating electrical signal. This applies to both EOM and MZM; The specific case of MZM is illustrated in Sect. C.1. The case for EAM follows a similar line of reasoning.

C.1 Mach Zehnder Interferometric Optical Modulator

Figure C.1 shows a schematic of a Mach Zehnder interferometric optical modulator fabricated as a guided-wave optical component. The optical waveguides are fabricated on an electro-optic material such that an applied voltage across the waveguide changes the effective refractive index seen by the propagating optical wave.

Assuming the modulator electrode impedance is matched to the microwave drive source, and that the traveling wave electrode is properly terminated so that no electrical reflection occurs at the end of the drive electrode, the (traveling wave) microwave drive signal along the electrode is $V(z,t) = V_0 \sin 2\pi f(n_m z/c - t)$. Where n_m is the microwave effective index of the traveling wave electrode, and f frequency of the modulation signal. If the

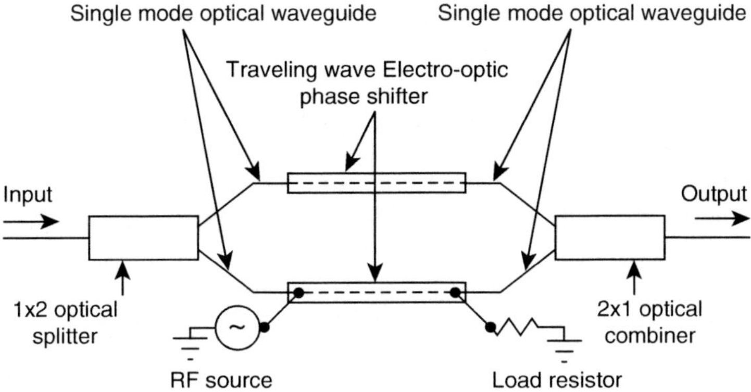

Fig. C.1. Schematic of a Mach Zehnder interferometric optical modulator fabricated as a guided-wave optical component

group index of the optical wave is n_0, then from the moving reference frame of the photons the voltage "seen" at position z is $V(z,t) = V_0 \sin 2\pi f(z n_m \delta/c - t)$ where $\delta = 1 - n_0/n_m$ is a velocity mismatch parameter between the microwave propagation along the drive electrode and the optical wave in the waveguide, and n_0 is the effective index of the optical waveguide. The phase modulation (in the case of Mach Zehnder interferometric modulator) or modulation in intensity attenuation (in the case of electroabsorption modulator) of the propagating optical wave is proportional, to lowest order, to the voltage "seen" locally by the propagating photon. At the exit of the modulated section of the waveguide the cumulative phase modulation for a photon entering the waveguide at $t = t_0$ (in the case of MZ modulator) is thus proportional to

$$\int_0^L \Delta\beta(f)\,dz = \frac{\overline{\Delta\beta}\sin\left(\dfrac{\pi f L n_m \delta}{c}\right)}{\left(\dfrac{\pi f L n_m \delta}{c}\right)} \sin\left(2\pi f t_0 - \frac{\pi f L n_m \delta}{c}\right), \qquad (C.1)$$

where $\overline{\Delta\beta} \sim V_0 L$ would be the maximum modulation in the phase shift at the output of the modulated waveguide section had there been no mismatch in the group velocities (index) of the optical and electrical waves, i.e., $\delta = 0$. For MZM's the optical wave from the modulated waveguide section interferes with the (unmodulated) wave from the other branch to produce an intensity modulation at the output of the device. Similar considerations apply for EAM's, resulting in a similar expression as (C.1) for traveling wave EAM's, with the optical phase shift $\Delta\beta(f)$ in (C.1) replaced by optical attenuation $\alpha(f)$. This optical phase modulation is converted into an optical intensity modulation after the modulated and original input waves are optically combined at the exit Y-branch of the MZ interferometer resulting in an optical intensity modulation proportional to $\sin(\Delta\beta)$.

C.2 Electro-Absorption Optical Modulator

Since the operation of the EAM relies on the shifting of the band edge in the semiconductor medium, the intrinsic modulation speed is limited only by how fast the electronic wavefunction of the bands can be modified. Similar to the case of electro-optic effect, for most practical purposes these time scales can be considered instantaneous. Which leaves the capacitance between the drive electrodes as the realistic limitation on how fast these optical modulators can practically operate. Section C.1 illustrated the use of a traveling wave – "TW" electrode structure to neutralize the parasitic capacitance associated with the drive electrode of a MZ modulator. A similar approach can be used for EAM's as well, as illustrated in Fig. C.2 which schematically depicts a "TW-EAM".

Using this approach, the only limitation to modulation speed is again due to the inevitable "velocity mismatch" between the optical wave in the optical

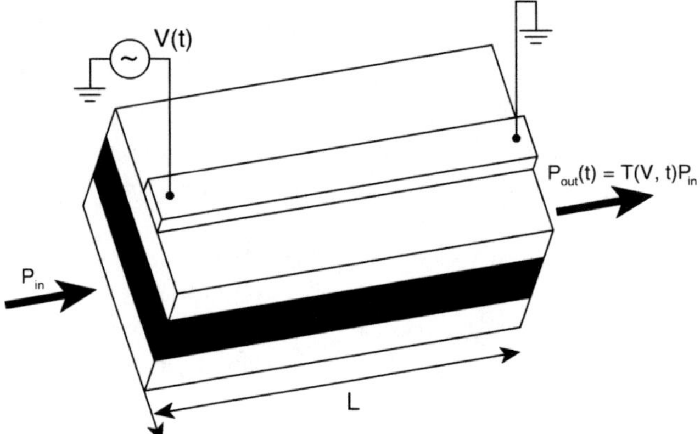

Fig. C.2. TW-EAM

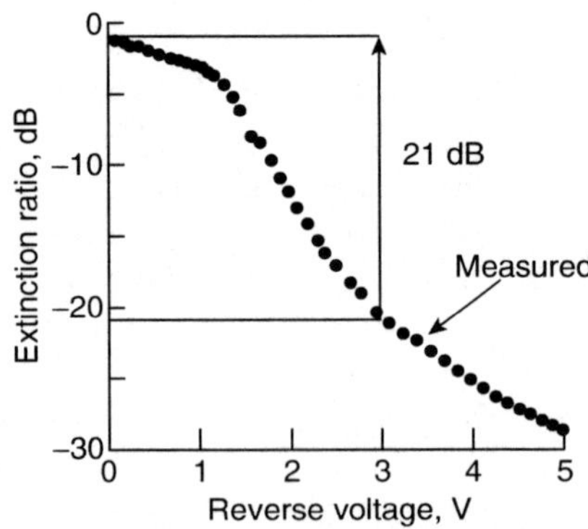

Fig. C.3. Measured transfer function. From [175]. ©2001 IEE. Reprinted with permission

waveguide and the electrical drive signal propagating along the drive electrode. The frequency response of a TW-EAM thus has the same form given by (C.1) in Sect. C.1 above.

Aside from modulation speed, linearity is also a consideration for applications contemplated in this book. While MZM modulators have a DC modulation transfer function (light-out vs. voltage-in) which is sinusoidal in nature, one can nevertheless operate the modulator at a bias point where the second order nonlinearity is minimized, i.e., at the inflexion points of the sine function. The modulation transfer function of a EOM is much less "ideal" in the sense that it cannot be represented by a simple function, with no clearly definable inflexion points.

A measured transfer function is shown in Fig. C.3. This is one of the reasons that EAM's are rarely used in linear lightwave transport systems. Implementations of fiber-optic transport of mm-wave signals using Mach-Zehnder electrooptic modulators as transmitters are described in Chaps. 12 and 13.

D

Modulation Response
of Superluminescent Lasers

D.1 Introduction

The superluminescent diode is one possible alternative to the well established injection laser's and LED's as light sources for fiber-optic communications. A superluminescent diode is basically a laser diode without mirrors. The first investigation of the superluminescent diode was carried out by Kurbatov et al. [176], its static properties were evaluated in detail by Lee et al. [177] and Amann et al. [178, 179]. Superluminescent diodes have also been integrated monolithically with detectors for optical memory readout, their fabrication being (possibly) simpler than that involved in laser-detector integration since a mirror facet is not required within the integrated device [180]. (However, given the high optical gain inside the device, it is not trivial how to eliminate even the minute amount of optical feedback to prevent lasing from occurring.) It has also been observed that [181, 182] the optical modulation bandwidth increases substantially as light emitting diodes enter the superluminescent regime. This regime is characterized by a rapid increase in the optical power output and a narrowing of the emission spectrum. This increase in modulation speed was attributed to the shortening of the carrier lifetime due to stimulated emission.

Due to the substantial non-uniformity in the longitudinal distribution of the photon and carrier densities in the active region, the modulation response of superluminescent. lasers cannot be described by the usual spatially uniform rate equations, which was so successful in describing laser dynamics. This is evident from the discussions in Chap. 1. Rather, the "local" rate equations of Chap. 1 should be used in their original form, necessitating numerical solutions of the coupled nonlinear space-time traveling wave rate equations. Results on numerical calculations of the small signal modulation frequency response will be described in this chapter. This exercise also presents an opportunity for determining the exact degree of validity of the spatial-average approximation used in the ubiquitous rate equations for studying modulation dynamics of semiconductor lasers. The results show that in most cases, the superluminescent diode modulation responses are of single-pole type, in contrast to the conjugate pole-pair response of a laser (Chap. 2). The cut-off frequency increases with pump current, similar to previous observations for a laser, albeit in a rather non-linear fashion. Under some conditions, the frequency response can be much higher than a laser diode of similar construction, under similar pump current densities. These conditions require that the reflectivity of the mirrors be less than 10^{-4}, and the spontaneous emission factor be less than 10^{-3}. The second of these conditions may be achieved with special device designs, but the first condition is challenging, as any minute imperfection in the waveguide construction and/or end facet elimination, or even feedback from the optical fiber into which the Superluminescent Laser couples can result in a retroreflection higher than that required to suppress lasing.

D.2 The Small Signal Superluminescent Equations and Numerical Results

The superluminescent diode is assumed to be constructed as a double heterostructure laser with guiding in both transverse directions, but with no end facet mirrors. The local rate equations for the photon and electron densities were first introduced in Chap. 1, (1.1). In the pure superluminescent case (no mirror), the steady state is given by the solutions (1.3)–(1.7) with $R = 0$.

Figure D.1a shows a plot of the steady state relative output optical power, $X_0^+(L/2) = X_0^-(-L/2)$, as a function of pumping level indicated by the unsaturated gain g, for various values of the spontaneous emission factor β. The linear part of the curves at the higher values of the pump level is the saturated regime, where most of the optical power is extracted from the inverted population by stimulated emission. Figure D.1b shows the static gain and photon distributions inside a superluminescent laser, illustrating the effect of spontaneous emission on the distributions.

To investigate the modulation frequency response of the superluminescent diode the usual perturbation expansion is employed

$$X^\pm(z,t) = X_0^\pm(z) + x^\pm(z)\mathrm{e}^{i\omega t} \tag{D.1a}$$

$$N(z,t) = N_0(z) + n(z)\mathrm{e}^{i\omega t} \tag{D.1b}$$

where $x^\pm$ and n are "small" variations about the steady states. This assumes that the electron and photon densities throughout the length of the diode vary in unison. This is true when propagation effects are not important, i.e., when modulation frequencies are small compared with the inverse of the photon transit time. This would amount to over 15 GHz even for very long diodes (0.25 cm) considered in later sections.

Adopting the usual technique of substituting (D.1) into the superluminescent equations (D.1), and neglecting the nonlinear product terms, the following small signal equations are obtained:

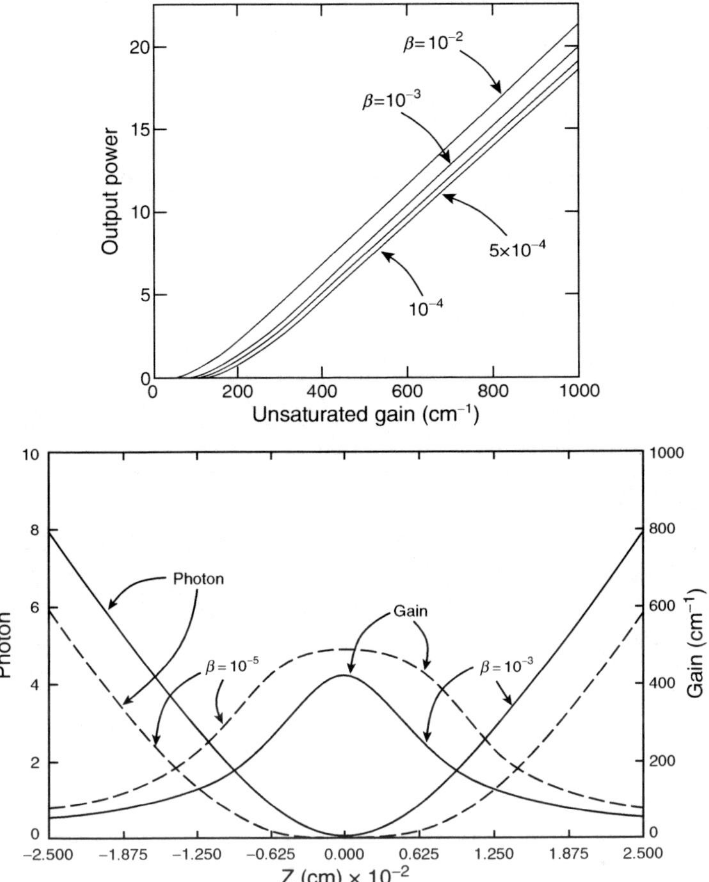

Fig. D.1. (**a**) Static photon output of, and (**b**) gain and photon distributions in a superluminescent diode. The unsaturated gain in (**b**) is $500\,\mathrm{cm}^{-1}$

$$\frac{dx^+}{dz} = Ax^+ + Bx^- + C \tag{D.2a}$$

$$\frac{dx^-}{dz} = Dx^+ + Ex^- + F \tag{D.2b}$$

where A, B, C, D, E, F are given by the following:

$$A = g_0 - \frac{i\omega}{c\tau_s} - \frac{(X_0^+ + \beta)g_0}{1 + i\omega + (X_0^+ + X_0^-)} \tag{D.3a}$$

$$B = \frac{-(X_0^+ + \beta)g_0}{1 + i\omega + (X_0^+ + X_0^-)} \tag{D.3b}$$

$$C = \frac{g_m(X_0^+ \beta)}{1 + i\omega + (X_0^+ + X_0^-)} \tag{D.3a}$$

$$D = \frac{(X_0^- + \beta)g_0}{1 + i\omega + (X_0^+ + X_0^-)} \tag{D.3b}$$

$$E = -\left(g_0 - \frac{i\omega}{c\tau_s} - \frac{(X_0^- + \beta)g_0}{1 + i\omega + (X_0^+ + X_0^-)}\right) \tag{D.3c}$$

$$F = \frac{-g_m(X_0^- + \beta)}{1 + i\omega + (X_0^+ + X_0^-)} \tag{D.3d}$$

where $g_0(z) = \alpha N_0(z) =$ small signal gain distribution, $g_m = \alpha j\tau_s/(ed) =$ small signal gain due to RF drive current, and ω has been normalized by the inverse of spontaneous lifetime.

The boundary conditions far solving (D.2) are the same as that in solving the steady state case, (1.2):

$$x^+(0) = x^-(0) \tag{D.4a}$$

$$x^-(L/s) = 0 = x^+(-L/2) \tag{D.4b}$$

Equation (D.2) is solved by assuming an arbitrary value for $x^+(0) = x^-(0) = \kappa$ and integrating (D.2) to give $x^+(L/2) = P$, $x^-(L/2) = Q$, these quantities are complex in general. The system (D.2) is integrated again assuming $x^+(0) = x^-(0) = \rho \neq \kappa$, to give $x^+(L/2) = T$, $x^-(L/2) = S$. The solution is given by a suitable linear combination of the above two solutions such that $x^-(L/2) = 0$. The small signal output of the superluminescent diode is

$$x^+\left(\frac{L}{2}\right) = x^-\left(-\frac{L}{2}\right) = \frac{QT - SP}{Q - S}. \tag{D.5}$$

The frequency response curve is obtained by solving (D.2) for each ω. One set of results is shown in Fig. D.2, which shows the frequency response of a $500\,\mu$n diode pumped to various levels of unsaturated gain. The spontaneous emission factor β is taken to be 10^{-4}, and the spontaneous lifetime 3 ns. The dashed curves are the phase responses. One noticeable feature is that the response is flat up to the fall-off frequency, and the fall-off is at approximately 10 dB/decade. The cut-off frequency (defined to be the abcissa of the intersection point between the high frequency asymptote and the 0 dB level of the amplitude response) easily exceeds 10 GHz at pump levels corresponding to unsaturated gain values of $1,000\,\text{cm}^{-1}$. This kind of gain may be unrealistic in real devices, but it will be shown further on that equally high frequency responses can be attained with longer devices at much lower pump levels.

In conventional (i.e., two-mirror) lasers, the spontaneous emission factor is found to play an important part in damping the resonance in the modulation response, while its effect on the corner frequency is not significant. In the case

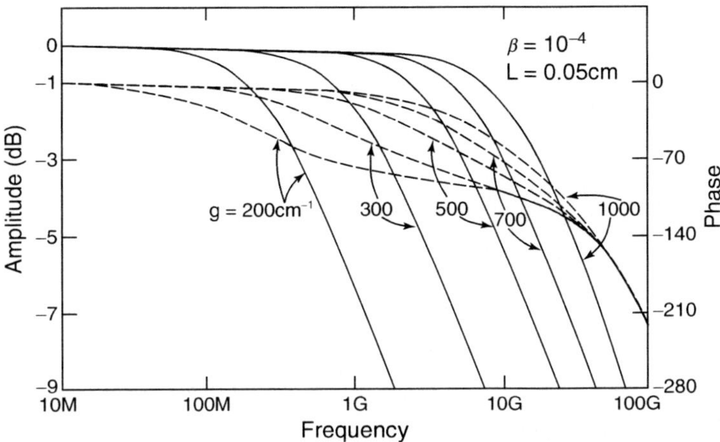

Fig. D.2. Amplitude and phase responses of a 500 μm superluminescent diode at various pumping levels. $\beta = 10^{-4}$

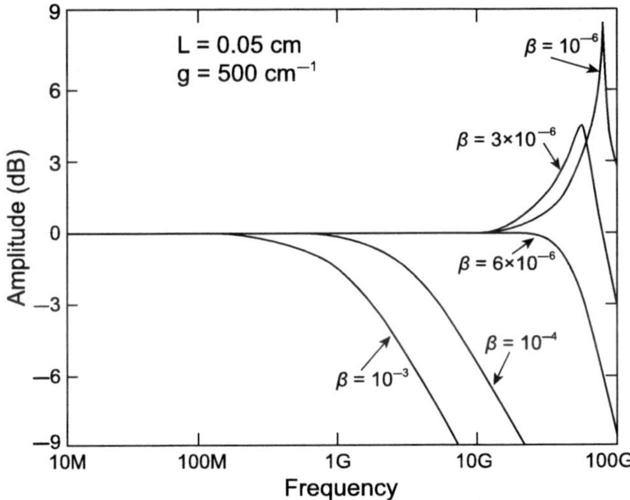

Fig. D.3. Amplitude response of a 500 μm diode with various β. Unsaturated gain = $500\,\mathrm{cm}^{-1}$

of superluminescent diodes, the spontaneous emission has a strong effect on both the damping as well as the magnitude of the resonance. This is shown in Fig. D.3, where one can plot the frequency response of a 500 μn diode pumped to an unsaturated gain of $500\,\mathrm{cm}^{-1}$, at different values of The modulation capability increases extremely fast as β is decreased, and at values below 5×10^{-6} a resonance peak appears and the fall-off approaches 20 dB/decade.

Figure D.4a shows plots of the comer frequency versus pumping, with different spontaneous emission factors β. The corner frequency increases rapidly

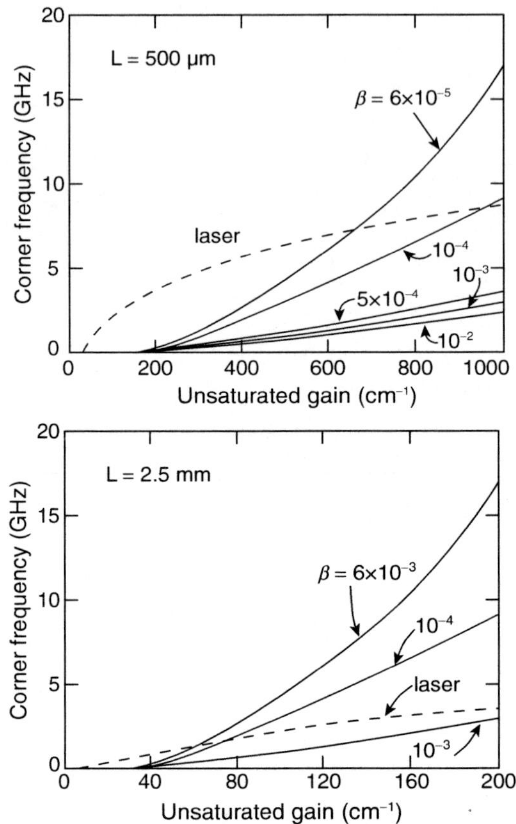

Fig. D.4. Corner frequencies of a superluminescent diode vs. pump, with various β, (above) for a 500 µm diode and (below) a 2.5 mm diode. The responses of conventional (0.3 mirror reflectivity) laser diodes of similar lengths are shown in *dotted lines*

as β falls below 10^{-3}. For comparison, the frequency response of a conventional laser with the same length of 500 µm, with an end mirror reflectivity of 0.3, is also shown in the figure. This curve is calculated using the well known formula (1) from [15]. It is apparent from Fig. D.4 (upper figure) that superluminescent diodes are not competitive with lasers of similar construction except at extremely high pump levels and for very small spontaneous emission factors. The spontaneous emission factor, depending on the waveguiding geometry, varies from about 10^{-5} in simple stripe geometry to 10^{-4} in lasers with real lateral guiding [183]. Since the superluminescent spectrum is at least an order of magnitude wider than the laser spectrum, the actual spontaneous emission factor should lie around 10^{-3}–10^{-4}. To obtain values of β as low as 10^{-6} would be difficult. However, one interesting feature of the frequency response of superluminescent diodes is that the corner frequency is invariant to gL, the unsaturated gain of the device. This is true for diode lengths as long as 0.3 cm, thus, the frequency response of a 0.2 cm diode pumped to an unsaturated gain of $200\,\mathrm{cm}^{-1}$ is identical to that of a $500\,\mu\mathrm{m}$ diode pumped to $800\,\mathrm{cm}^{-1}$ – an unrealistic value. Conventional lasers do not have this property since longer diodes have a longer photon lifetime, and would have a lower corner frequency when pumped at the same level of g/g_{th}. In other words, very high frequency responses can be attained at very modest pump current densities by using very long diodes, as illustrated in Fig. D.4 (lower figure), which is a similar plot to Fig. D.4 (upper figure) but for a 0.25 cm diode.

D.3 Effect of a Small but Finite Mirror Reflectivity

In practice, the reflection from mirror facets cannot be reduced to absolute zero. By using Lee's structure [177] or by placing the waveguide at an angle to the facets, the only feedback into the waveguide mode is due to scattering, which can be made very small. However, even with a reflectivity as small as 10^{-6}, the conventional threshold gain (neglecting internal absorption loss) of a $500\,\mu\mathrm{m}$ laser is $\frac{1}{L}\ln\frac{1}{R} \approx 276\,\mathrm{cm}^{-1}$; for longer diodes the threshold decreases inversely to L. Thus at pumping levels of interest ($gL > 20$) the diode can well be above the conventional lasing threshold. The question arises as to whether the frequency response of such a device behaves like that of a conventional laser diode, with a square root dependence as in (1) from [15], or like that of a superluminescent diode. There is actually no reason to believe that a laser with mirror reflectivities as low as 10^{-6} should behave as predicted by the spatially uniform rate equations. The optical spectrum, though wider than common laser diodes because of the extremely low finesse cavity, would be considerably narrower than the free superluminescent spectrum. The following calculations are carried out to illustrate the effects of a small but finite mirror reflectivity on the frequency response of the superluminescent diode. *It also serves to illustrate the actual range of validity of the conventional rate equations.* In the case of a finite reflectivity the pertinent boundary conditions are

$$x^{-}(L/2) = Rx^{+}(L/2) \tag{D.6a}$$
$$x^{+}(-L/2) = Rx^{-}(-L/2) \tag{D.6b}$$

where R is the mirror reflectivity. The above boundary conditions modify (D.5) into

$$x^{+}(L/2) = x^{-}(-L/2) = \frac{QT - SP}{(Q - P) + R(T - P)}. \tag{D.7}$$

A set of results is shown in Figs. D.5 where the frequency response is shown for a long (0.25 cm) diode with $\beta = 10^{-3}$ and 10^{-4}, and assuming a mirror reflectivity of 10^{-6}. Features of both the laser diode and the superluminescent diode can be observed in the frequency response. The corner frequency is sensitive to the spontaneous emission factor, increases much faster than a

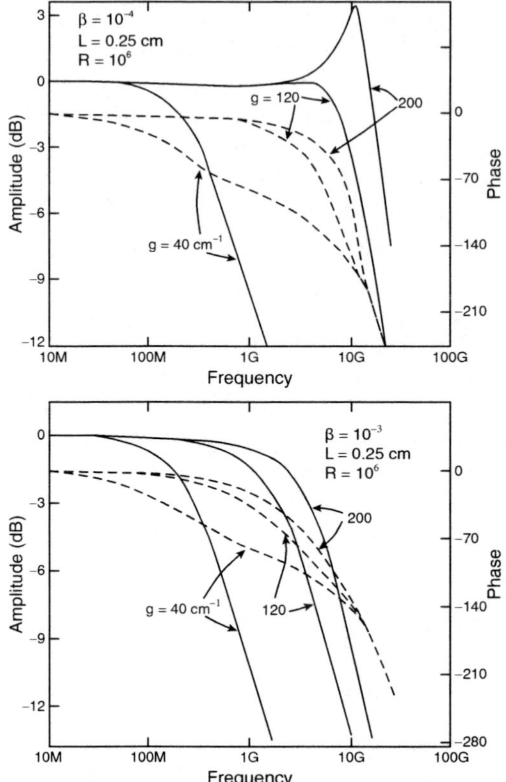

Fig. D.5. Frequency response of a 0.25 cm diode with mirror reflectivity of 10^{-6}, and $\beta = 10^{-4}$ in (top) and 10^{-3} in (bottom). g is the unsaturated gain in cm^{-1}

square root dependence on pumping, but at a sufficiently high pump level a resonance peak occurs similar to a conventional laser.

Figure D.6 shows plots of the corner frequency versus pump level, for spontaneous emission factors of 10^{-3} and 10^{-4}. Also shown is the response of a laser of similar length (0.25 cm) with a reflectivity of 0.3. As the mirror reflectivity increases from 10^{-6}, the superluminescent response curve merges continuously onto the laser curve and becomes essentially the laser curve at reflectivities around 10^{-3}.

Using a plausibility argument one can formulate a rough criterion as to how small the end mirror reflectivity should be for the frequency response to be super-luminescent-like. In the pure superluminescent case the photons are generated at one end through spontaneous emission, and are amplified as they traverse the active medium. Thus, if the product of the reflectivity and the outward traveling photon density at a mirror facet is larger than β, then the device resembles more closely a conventional laser than an amplified spontaneous emission device. With $\beta = 10^{-4}$ and a typical normalized photon

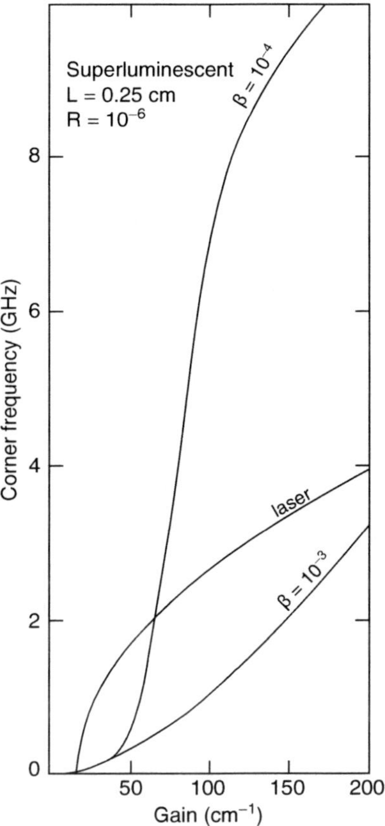

Fig. D.6. Corner frequency of a 0.25 cm diode with mirror reflectivity of 10^{-6}, at spontaneous emission factors of 10^{-4} and 10^{-3}. The response of a conventional 0.25 cm laser is also shown

density of 10 near the output end, the mirror reflectivity must be smaller than 10^{-5} for the response to be superluminescent-like. This plausibility argument is supported by results of numerical calculations.

Finally, for longer devices one cannot neglect internal optical loss, which amounts to about $10\text{--}20\,\mathrm{cm}^{-1}$ [29] in common GaAs laser materials. The optical output power will not increase linearly as shown in Fig. D.1a, but will saturate at a value X_g, given by

$$\frac{g}{1 + X_g} = f \, , \tag{D.8}$$

where f is the internal loss in cm^{-1}, g is the unsaturated gain. For devices as long as 0.25 cm, the Internal loss is considerably lower than the saturated gain anywhere inside the active medium except for a small region near the ends, where the photon density is highest. The effect on the frequency response proves to be insignificant.

Broadband Microwave Fiber-Optic Links with RF Phase Control for Phased-Array Antennas

Use of optical fiber for the efficient transport of microwave signals in broadband phased-array antenna systems has been the subject of intense interest for the last several years [184–186]. Although the advantages of a phased-array system based on true-time-delay is well known, a means of providing continuous RF phase adjustment to complement the discrete, fixed fiber-delay lines is needed for the practical implementation of precise antenna beam steering. This is currently done with electronic phase shifters at the antenna site [186]. In this appendix, it is shown that by injection locking the self-pulsating frequency of a commercially available self-pulsing semiconductor laser diode (SP-LD) – an inexpensive CD laser diode emitting at $\sim$850 nm, an optical transmitter can be built in which the RF phase of the modulated optical output can be *continuously* adjusted by varying the bias current into the laser. Phase-shifting over a range of 180° with a switching time of <5 ns can be achieved. The SP-LD phase shifter can operate over a bandwidth of >7 GHz by injection locking higher harmonics of the self-pulsation. The principle of operation is based on the characteristics of a conventional electronic injection-locked oscillator. It is well known that by detuning the free-running frequency of an injection-locked oscillator from the injected signal frequency, the phase of the locked output signal can be varied between $\pm 90°$ within the locking bandwidth [187]. In this case, the SP-LD acts as a frequency tunable, free-running oscillator whose output is in optical form, and whose frequency can be tuned over an exceptionally wide range of 1–7 GHz. Phase shifting of RF phase in the modulated optical output is conveniently done by varying the free-running self-pulsing frequency. The characteristics of the laser used is described in [188,189]. To illustrate locking of the self-pulsation by an external RF source, Fig. E.1a shows the measured RF spectrum of the free-running and locked signal at the edge of the locking band for an injected signal frequency $f_{\mathrm{inj}} = 1.3$ GHz and an injected power $P_{\mathrm{inj}} = 0$ dBm.

A "strong" lock is clearly evident. Figure E.1b shows the measured phase shift of the input RF carrier in the time-domain, where the bias current of the SP-LD is varied by ~ 1 mA to produce a phase shift of 60°. The setup

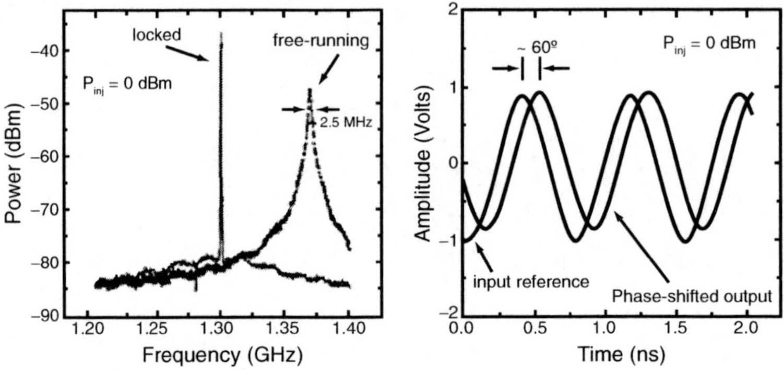

Fig. E.1. (a) Injection-locked and free-running spectrum at edge of locking band. (b) Measured steady-state phase shift in the time domain for $\Delta I_{dc} \approx 1\,\mathrm{mA}$

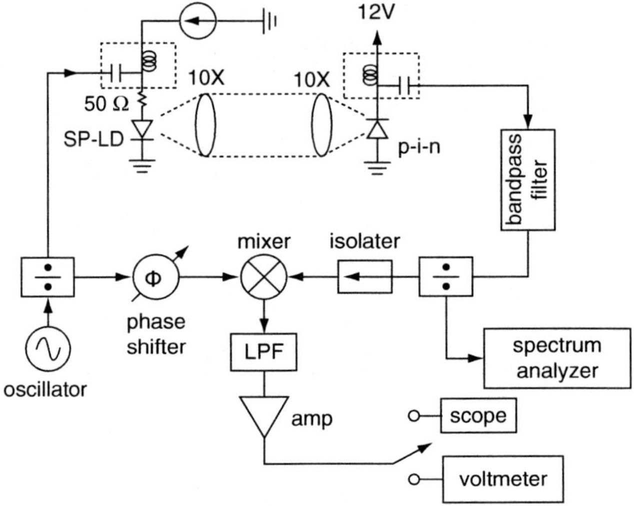

Fig. E.2. Setup used to measure the phase as a function of bias current and the pulse response of the phase shifter

illustrated in Fig. E.2 was used to measure the phase as a function of the bias current for various injected signal powers.

A portion of the driving RF carrier is fed into the mixer LO port and serves as the reference signal. The intensity coming from the injection-locked SP-LD is collimated and focused onto a standard high-speed photodetector. A portion of the detected signal is displayed on a RF spectrum analyzer to monitor the carrier-to-noise ratio (CNR). Note that it is *not* necessary to place a microwave amplifier after the photodetector to achieve an overall net electrical gain for the link. It will be shown below that it is possible for the SP-LD phase shifter/optical link to provide an inherent link gain of up to +10 dB. When the bias current into the SP-LD is varied the resulting RF output phase

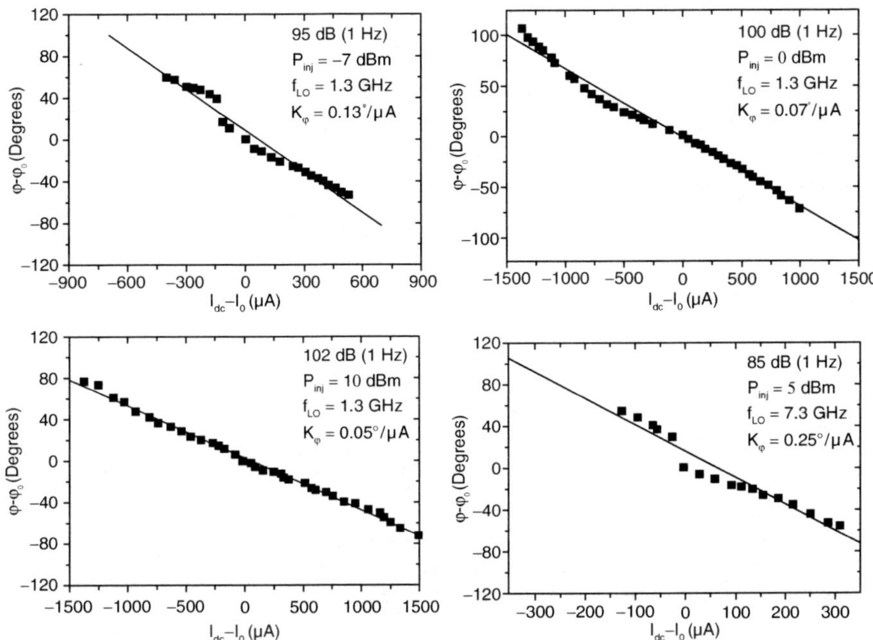

Fig. E.3. Measured phase shift as a function of bias current (**a**) $P_{\text{inj}} = -7\,\text{dBm}$, $B_L \approx 80\,\text{MHz}$. (**b**) $P_{\text{inj}} = 0\,\text{dBm}$, $B_L \approx 170\,\text{MHz}$. (**c**) $P_{\text{inj}} = 10\,\text{dBm}$ and $B_L > 200\,\text{MHz}$ (outside the filter bandwidth). (**d**) Injection-locking to third harmonic of $f_{sp} = 2.43\,\text{GHz}$; $B_L \approx 120\,\text{MHz}$

shift can be observed by measuring the DC voltage at the output of the mixer. The results are displayed in Fig. E.3 for various injected signal powers.

As expected, the CNR (measured at 1 MHz offset) is increased by increasing the injected signal power. A linear regression fit is performed for all plots. In Fig. E.3b the injected signal power is 0 dBm. The phase shift covers a range of 180° (from -70 to $110°$), and control of the bias current to within $\sim 10\,\mu\text{A}$ allows a phase resolution of $< 1°$ (i.e., $[K_\phi = 0.07°/\mu\text{A}]\,[10\,\mu\text{A}] = 0.7°$). The maximum phase deviation from the linear regression fit is $10°$. A maximum CNR of 102 dB (1 Hz) is obtained at an injected power of 4 dBm or more. This is displayed in Fig. E.3c, where the injected microwave signal power is set to 10 dB. Note that for this plot the phase resolution is $0.5°$ (for a $10\,\mu\text{A}$ change in the bias current). The deviation from linearity is $< 5°$. The slope (K_ϕ) for each graph is related to locking bandwidth B_L, and the locking bandwidth is a function of the injected signal power P_{inj}

$$\frac{1}{K_\phi} \propto B_L \propto \sqrt{P_{\text{inj}}} \tag{E.1}$$

As mentioned above, the injected power also determines the CNR. Detuning the self-pulsing frequency from the injected signal frequency produces the desired phase shift, but also decreases the CNR. The locking bandwidth is defined as the detuning in which the CNR is maintained to within 3 dB of the maximum (which occurs at $f_{\mathrm{inj}} = f_{sp}$) at a given injected RF power. The locking bandwidth for each plot is given in the caption for Fig. E.3. Figure E.3d shows locking of the third harmonic whose fundamental $f_{sp} = 2.43\,\mathrm{GHz}$. This increases the operational range of the phase shifter/optical link to $>7\,\mathrm{GHz}$. Although the injected power noted in this plot is $5\,\mathrm{dBm}$, the actual injected microwave power into the laser is considerably less, as these commercial compact disk lasers are not packaged for high frequency operation. Commercial SP-LD's pulsate at a fundamental frequency up to $5\,\mathrm{GHz}$ [188, 189], and can be extended to beyond $7\,\mathrm{GHz}$ by appropriate design [190]. Thus, it is possible that by injection locking to the fourth or fifth harmonic, operation can reach into the millimeter-wave region.

The step response of the phase shifter was also measured. A 500-kHz square wave with a rise and fall time of 3 ns and peak-to-peak amplitude of 200 mV was superimposed on top of the RF signal, which drives the SP-LD. The square wave acts as the phase-control signal to the SP-LD phase shifter, and its amplitude determines the resulting phase of the RF signal at the output of the link. For a peak-to-peak amplitude of 200 mV, the phase change is $\Delta\phi = K_\phi(V_{pp}/R) = (0.07°/\mu\mathrm{A})(200\,\mathrm{mV}/50\,\Omega) = 140°$. The injection-locked phase-modulated carrier is detected at the output of the mixer on a sampling oscilloscope. Figure E.4 shows the measured pulse response. The rise time is $<5\,\mathrm{ns}$ for a phase transition of $\approx 140°$. The fall time was also measured to be $5\,\mathrm{ns}$.

A feature of the injection-locked SP-LD phase shifter/optical link is that the net electrical gain of the link can exceed unity, without the use of a microwave preamplifier. For ideal (lossless) impedance matching of the laser transmitter and receiver, the link gain G_L is given by [191]

$$G_L(P_{\mathrm{inj}}) = \frac{P_{out}}{P_{\mathrm{inj}}} = \frac{[\eta_L K_{\mathrm{opt}} \eta_D]^2}{4}\left[\frac{R_p}{R_d}\right] G_{\mathrm{inj}}(P_{\mathrm{inj}}),\qquad\text{(E.2)}$$

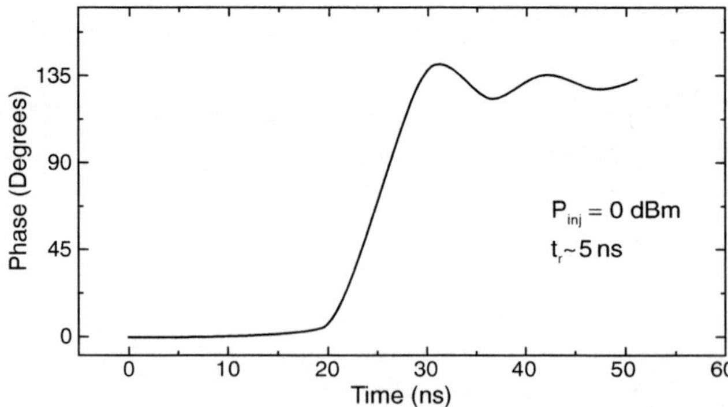

Fig. E.4. Step response of injection-locked SP-LD phase shifter

Table E.1. Comparison of a typical electronic phase shifter to the SP-LD phase shifter

Parameter	SP-LD phase shifter	Electronic phase shifter
Phase range	$180°$	$360°$
Gain (loss)	$\geq 10\,\mathrm{dB}$	$\geq -1\,\mathrm{dB}$
Isolation	$>80\,\mathrm{dB}$	$<40\,\mathrm{dB}$
Bandwidth	$>7\,\mathrm{GHz}$	$<5\,\mathrm{GHz}$
Resoluation	$\sim 1°$	$\sim 5°–10°$
Switching time	$5\,\mathrm{ns}$	$>10\,\mathrm{ns}$

where η_L, η_D are the laser and detector responsivity; and K_{opt} includes all of the optical coupling losses between the laser and the detector. G_{inj} is the gain due to the injection-locking process (as compared to a conventional directly modulated optical link), and is a function of the injected signal power. For typical values of $\eta_L = 0.35\,\mathrm{W/A}$, $\eta_D = 0.6\,\mathrm{A/W}$, $K_{\mathrm{opt}} = 0.90\,\mathrm{W/W}$, a photodiode impedance of $R_p = 1\,\mathrm{k\Omega}$, a laser diode forward resistance $R_d = 5\,\Omega$, and a measured $G_{\mathrm{inj}}(P_{\mathrm{inj}} = 0\,\mathrm{dB}) = 5$, an overall net link gain of $10\,\mathrm{dB}$ is obtained. This is a desirable feature since conventional, passive electronic phase shifters will introduce a net RF loss of anywhere from $1–15\,\mathrm{dB}$. The RF isolation is also $>80\,\mathrm{dB}$ for the optical link [191]. On the other hand, a disadvantage of the present approach is the relatively high level of (phase) noise compared to directly modulated optical links that use conventional, passive electronic phase shifters. This arises from the inherent noise associated with an active oscillator, even under injection-locked conditions. Whether this results in a performance compromise at the system level depends on the system architecture. Apart from this issue, the phase-shifting performance of the SP-LD phase shifter with that of a conventional electronic phase shifter is compared in Table E.1.

It is shown that an injection-locked self-pulsating laser can be used as a continuously adjustable phase shifter/optical transmitter for optical control of phased-array antenna, operating in conjunction with fiber delay lines. They compare favorably with conventional optical links incorporating conventional electronic phase-shifters in aspects of operational frequency range, switching speed, RF link gain, and isolation. It is important to note that these SP-LD lasers are low-cost and commercially available. A higher link noise due to the use of an active oscillator should be considered along with the above advantage in overall system design.

References

1. W.I. Way, *Broadband Hybrid Fiber/Coax Access System Technologies.* (Academic, New York, 1998)
2. J.B. Moreno, J. Appl. Phys. **48**(10), 4152 (1977)
3. W.E. Lamb Jr, Phys. Rev. **134**(6A), 1429 (1964); A. Icsevgi, W.E. Lamb Jr, *Phys. Rev.* **185**(2), 517 (1969)
4. L.W. Casperson, *J. Appl. Phys.* **46**(12), 5194 (1975)
5. F. Stern, *IEEE J. Quantum Electron.* **9**(2), 290 (1973).
6. H. Kressel, J.K. Butler, *Semiconductor Lasers and Heterojunction LEDs.* (Academic, New York, 1977), p. 77
7. R.H. Pantell, H.E. Puthoff, *Fundamentals of Quantum Electronics.* (Wiley, New York, 1969), p. 294
8. N. Chinone, K. Aiki, M. Nakamura, *IEEE J. Quantum Electron.* **14**(8), 625 (1978)
9. D.P. Wilt, A. Yariv, *IEEE J. Quantum Electron.* **17**(9), 1941 (1981)
10. D.A. Kleinman, *Bell Syst. Tech. J.* **43**, 1505 (1964)
11. R. Salathe, C. Voumard, H. Weber, *Opto-Electron.* **6**(6), 451 (1974)
12. L. Figueroa, C. Slayman, H.W. Yen, *IEEE J. Quantum Electron.* **18**(10), 1718 (1982)
13. K.Y. Lau, A. Yariv, *Appl. Phys. Lett.* **40**(6), 452 (1982)
14. F. Stern, *J. Appl. Phys.* **47**(12), 5382 (1976)
15. K.Y. Lau, N. Bar-Chaim, I. Ury, C. Harder, A. Yariv, *Appl. Phys. Lett.* **43**(1), 1 (1983)
16. N. Bar-Chaim, J. Katz, I. Ury, A. Yariv, *Electron. Lett.* **17**(3), 108 (1981)
17. I. Ury, K.Y. Lau, N. Bar-Chaim, A. Yariv, *Appl. Phys. Lett.* **41**(2), 126 (1982)
18. Hitachi laser diode application manual
19. G. Arnold, P. Russer, *Appl. Phys. Lett.* **14**, 255 (1977)
20. T. Ikegami, Y. Suematsu, *Electron. Commun. Jpn.* **53**(9), 69 (1970)
21. T.H. Hong, Y. Suematsu, *Trans. Inst. Electron. Commun. Eng. Jpn. E (English)* **E62**(3), 142 (1979)
22. K.E. Stubkjr, *Electron. Lett.* **15**(2), 61 (1979)
23. K.Y. Lau, A. Yariv, *Opt. Commun.* **34**(3), 424 (1980)
24. K. Otsuka, *IEEE J. Quantum Electron.* **13**, 520 (1977)
25. M. Nagano, K. Kasahara, *IEEE J. Quantum Electron.* **13**, 632 (1977)
26. K.Y. Lau, C. Harder, A. Yariv, *Appl. Phys. Lett.* **44**(3), 273 (1984)

27. K.Y. Lau, N. Bar-Chaim, I. Ury, A. Yariv, *Appl. Phys. Lett.* **45**(4), 316 (1984)

28. K.Y. Lau, A. Yariv, *Appl. Phys. Lett.* **45**(10), 1034 (1984)

29. H. Kressel, J.K. Butler, *Semiconductor lasers and heterojunction LEDs*, (Academic, New York, 1979)

30. H. Blauvelt, S. Margalit, A. Yariv, *Appl. Phys. Lett.* **40**(12), 1029 (1982)

31. S. Takahashi, T. Kobayashi, H. Saito, Y. Furukawa, *Jpn. J. Appl. Phys.* **17**(5), 865 (1978)

32. K.Y. Lau, N. Bar-Chaim, I. Ury, Ch. Harder, A. Yariv, *Appl. Phys. Lett.* **43**(4), 329 (1983)

33. N. Bar-Chaim, K.Y. Lau, I. Ury, A. Yariv, *Appl. Phys. Lett.* **43**(3), 261 (1983)

34. M.A. Newkirk, K.J. Vahala, *Appl. Phys. Lett.* **54**(7), 600 (1988).

35. H.C. Casey, M.B. Panish, *Heterostructure Lasers.* (Academic, New York, 1978), Pt. A, pp. 174

36. K.Y. Lau, A. Yariv, *Semiconductor and Semimetals.* (Academic, New York, 1985)

37. G. Liu, S.L. Chuang, *IEEE J. Quantum Electron.* **37**(10), 1283 (2001)

38. Y. Arakawa, K. Vahala, A. Yariv, *Appl. Phys. Lett.* **45**(9), 950 (1984)

39. P. Zory, *Quantum well lasers.* (Academic, New York, 1993)

40. Y. Arakawa, H. Vahala, A. Yariv, K.Y. Lau, *Appl. Phys. Lett.* **47**(11), 1142 (1985)

41. The number of longitudinal modes depends, among other things, on the spontaneous emission factor; see D. Renner, J.E. Carroll, *Electron. Lett.* **14**, 781 (1978)

42. Gain guided lasers exhibit multilongitudinal mode oscillation due to a large spontaneous emission factor; see W. Streifer, D.R. Scifres, R.D. Burnham, *Appl. Phys. Lett.* **40**, 305 (1982)

43. C.L. Tang, H. Statz, G. DeMars, *J. Appl. Phys.* **34**, 2289 (1963)

44. K. Petermann, *Opt. Quantum Electron.* 10, 233 (1978)

45. M.R. Matthews, A.G. Steventon, *Electron. Lett.* **14**, 649 (1978)

46. F. Mengel, V. Ostoich, *IEEE J. Quantum Electron.* **13**, 359 (1977)

47. P.R. Seeway, A.R. Goodwin, *Electron. Lett.* 12, 25 (1976)

48. T.P. Lee, C.A. Burrus, P.L. Liu, A.G. Dentai, *Electron. Lett.* **18**, 805 (1982)

49. T.P. Lee, C.A. Burrus, R.A. Linke, R.J. Nelson, *Electron. Lett.* **19**, 82 (1983)

50. P.L. Liu, T.P. Lee, C.A. Burrus, I.P. Kaminow, J.S. Ko, *Electron. Lett.* **18**, 904 (1982)

51. G.H.B. Thompson, *Physics of semiconductor laser devices.* (Wiley, New York, 1980), p. 450

52. M. Nakamura, K. Aiki, N. Chinone, R. Ito, J. Umeda, *J. Appl. Phys.* **49**, 4644 (1978)

53. K. Kishino, S. Aoki, Y. Suematsu, *IEEE J. Quantum Electron.* **18**, 343 (1982)

54. Y. Sakakibara, K. Furuya, K. Utaka, Y. Suematsu, *Electron. Lett.* **16**, 456 (1980)

55. K. Utaka, I. Kobayashi, Y. Suematsu, *IEEE J. Quantum Electron.* **17**, 651 (1981)

56. K.J. Ebeling, L.A. Coldren, B.I. Miller, J.A. Rentschler, *Appl. Phys. Lett.* **42**, 6 (1983)

57. J. Buus, M. Danielsen, *IEEE J. Quantum Electron.* **13**, 669 (1977)

58. D. Psaltis, private communication

59. M. Yamada, Y. Suematsu, *J. Appl. Phys.* **52**, 2653 (1981)

60. T.L. Koch, J.E. Bowers, *Electron. Lett.* **20**, 1038 (1984)
61. P.J. Corvini, T.L. Koch, *IEEE J. Lightwave Tech.* **LT-5**, 1591 (1987)
62. L.F. Lester, S.S. O'Keefe, W.J. Schaff, L.F. Eastman, *Electron. Lett.* **28**(4), 383 (1992)
63. R. Nagarajan, T. Fukushima, J.E. Bowers, R.S. Geels, L.A. Coldren, *Appl. Phys. Lett.* **58**(21), 2326 (1991)
64. K. Uomi, H. Nakono, N. Chinone, *Electron. Lett.* **25**(10), 668 (1989)
65. K. Petermann, *Laser Diode Modulation and Noise.* (Kluwer, New York, 1988)
66. M. Yamada, *IEEE J. Quantum Electron.* **22**(7), 1052 (1986)
67. G.J. Meslener, *IEEE Photon. Tech. Lett.* **4**(8), 939 (1992)
68. K.Y. Lau, H. Blauvelt, *Appl. Phys. Lett.* **52**(9), 694 (1988)
69. M.M. Choy, J.L. Gimlett, R. Welter, L.G. Kazovsky, N.K. Cheung, *Electron. Lett.* **23**(21), 1151 (1987)
70. J.L. Gimlett, N.K. Cheung, *J. Lightwave Tech.* **7**(6), 888 (1989)
71. S. Wu, A. Yariv, H. Blauvelt, N. Kwong, *Appl. Phys. Lett.* **59**(10), 1156 (1991)
72. R.H. Wentworth, G.E. Bodeep, T.E. Darcie, *J. Lightwave Tech.* **10**(1), 84 (1992)
73. C.B. Su, J. Schlafer, R.B. Lauer, *Appl. Phys. Lett.* **57**(9), 849 (1990)
74. K.Y. Lau, A. Yariv, *Appl. Phys. Lett.* **47**(2), 84 (1985)
75. P.K. Pepeljugoski, K.Y. Lau, *J. Lightwave Tech.* **10**(7), 957 (1992)
76. A. Yariv, H. Blauvelt, S.-W. Wu, *J. Lightwave Tech.* **10**(7), 978 (1992)
77. K. Vahala, A. Yariv, *IEEE J. Quantum Electron.* **19**(6), 1096 (1983)
78. C. Harder, J. Katz, S. Margalit, J. Shacham, A. Yariv, *IEEE J. Quantum Electron.* **18**(3), 333 (1982)
79. C.H. Henry, *J. Lightwave Tech.* **4**(3), 288 (1986)
80. R.C. Alferness, G. Eisenstein, S.K. Korotky, R.S. Tucker, L.L. Buhl, I.P. Kaminow, J.J Veselka, paper WJ3, *Optical Fiber Communication Conference,* New Orleans, 1984
81. R.S. Tucker, G. Eisenstein, I.P. Kaminow, *Electron. Lett.* **19**(14), 552 (1983)
82. P.T. Ho, L.A. Glasser, E.P. Ippen, H.A. Haus, *Appl. Phys. Lett.* **33**, 241 (1978)
83. K.Y. Lau, A. Yariv, *Appl. Phys. Lett.* **46**(4), 326 (1985)
84. L. Figueroa, K.Y. Lau, H.W. Yen, A. Yariv, *J. Appl. Phys.* **51**(6), 3062 (1980)
85. J.E. Bower, C.A. Burrus, paper M-1, *Ninth International Semiconductor Laser Conference,* Rio de Janeiro, 1984
86. R.A. Linke, paper M-4, *Ninth International Semiconductor Laser Conference,* Rio de Janeiro, 1984
87. J.P. Van der Ziel, *Semiconductor and Semimetals,* vol. 22, part B, chap. 1, and references therein. (Academic, New York)
88. K.Y Lau, *Appl. Phys. Lett.* **52**(26), 2214 (1988)
89. S. Akiba, G.E. Williams, H.A. Haus, *Eletron. Lett.* **17**(15), 527 (1981)
90. G. Eisenstein, R.S. Tucker, U. Koren, S.K. Korotky, *IEEE J. Quantum Electron.* **22**(1), 142 (1986)
91. S.W. Corzine, J.E. Bowers, G. Przybylek, U. Koren, B.I. Miller, C.E. Soccolich, *Appl. Phys. Lett.* **52**(5), 348 (1988)
92. K.Y. Lau, I. Ury, A. Yariv, *Appl. Phys. Lett.* **46**(12), 1117 (1985)
93. J. Au Yeung, *IEEE J. Quantum Electron.* **17**(3), 398 (1981)
94. H.A Haus, *IEEE J. Quantum Electron.* **11**(7): 323 (1975)
95. A.E. Siegman, *Lasers.* (University Science Books, Mill Valley, CA, 1986)
96. H.A. Haus, *IEEE J. Quantum Electron.* **11**(9), 736 (1975)

97. H.A. Haus, *J. Appl. Phys.* **46**(7), 3049 (1975)

98. H.A. Haus, *IEEE J. Quantum Electron.* **12**(3), 169 (1976)

99. K.Y. Lau, P.L. Derry, A. Yariv, *Appl. Phys. Lett.* **52**(2), 88 (1988)

100. M.C. Wu, Y.K. Chen, T. Tanbun-Ek, R.A. Logan, M.A. Chin, *Appl. Phys. Lett.* **57**, 759 (1990)

101. H. Ogawa, D. Polifko, S. Banba, *IEEE Trans. Microw. Theory Tech.* **40**(12), 2285 (1992)

102. J. O'reilly, P. Lane, *J. Lightwave Tech.* **12**(2), 369 (1994)

103. J.B. Georges, M.-H. Kiang, K. Heppell, M. Sayed, K.Y. Lau, *IEEE Photon. Tech. Lett.* **6**(4), 568 (1994)

104. W.I. Way, *IEICE Trans. Commun.* **E76-B**(9), 1091 (1993)

105. J.B. Georges, D.M. Cutrer, M.-H Kiang, K.Y. Lau *IEEE Photon. Tech. Lett.* **7**(4), 431 (1995)

106. J.B. Georges, T.C. Wu, D.M. Cutrer, U. Koren, T.L. Koch, K.Y. Lau, "Millimeter-wave optical transmitter at 45 GHz by resonant modulation of a monolithic tunable DBR laser," *CLEO '95, Summaries of Papers Presented at the Conference on Lasers and Electro-Optics (IEEE Cat. No. 95CH35800)*, pp. 337-338, Opt Soc America, 1995, Washington, DC, USA

107. A.S. Daryoush, *IEEE Trans. Microw. Theory Tech.* **38**(5), 467 (1990)

108. D.A. Tauber, R. Spickermann, R. Nagarajan, T. Reynolds, A.L. Holmes Jr., J.E. Bowers, *Appl. Phys. Lett.* **64**(13), 1610 (1994)

109. D.M. Cutrer, J.B. Georges, T.-C. Wu, B. Wu, K.Y. Lau, *Appl. Phys. Lett.* **66**(17), 2153 (1995)

110. K.Y. Lau, *IEEE J. Quantum Electron.* **26**(2), 250 (1990)

111. S. Ramo, J.R. Whinnery, T. Van Duzer, *Fields and waves in communication electronics.* (Wiley, New York, 1965)

112. J. Park, K.Y. Lau, *Electron. Lett.* **32**(5), 474 (1996)

113. G.J. Meslener, *IEEE J. Quantum Electron.* **20**(10), 1208 (1984)

114. H. Schmuck, *Electron. Lett.* **31**(21), 1848 (1995)

115. C.S. Oh, W. Gu, *IEEE J. Sel. Area. Commun.* **8**(7), 1296 (1990)

116. A.F. Elrefaie, C. Lin, *Proceedings IEEE Symposium on Computers and Communications (Cat. No.95TH8054)*, IEEE Comput Soc Press, 1995, pp. 328–337, Los Alamitos, CA, USA

117. J. Park, A.F. Elrefaie, K.Y. Lau, *IEEE Photon. Tech. Lett.* **8**(12), 1716 (1996)

118. Cadence Releases Version 4.6 of Signal Processing Worksystem with New Libraries and System C 1.0 Co-Simulation Capability, http://www.cadence.com/company/newsroom/press_releases/pr.aspx?xml = 013101_SPW, retrieved: 2008/04/08, 12:48AM

119. R. Hofstetter, H. Schmuck, R. Heidemann, *IEEE Trans. Microw. Theory Tech.* **43**(9,pt. 2), 2263 (1995)

120. D.W. Dolfi, T.R. Ranganath, *Electron. Lett.* **28**(13), 1197 (1992)

121. D.A. Atlas, R.E. Pidgeon, D.W. Hess, "Clipping limit in externally modulated lightwave CATV systems," *OFC'96, Optical Fiber Communication*, vol. 2, 1996 Technical Digest Series, Conference Edition (IEEE Cat. No.96CH35901), Opt Soc America, pp. 282–283, Washington, DC, USA

122. M.R. Phillips, T.E. Darcie, D. Marcuse, G.E. Bodeep, N.J. Frigo, *IEEE Photon. Tech. Lett.* **3**(5), 481–483 (1991)

123. I.M.I. Habbab, A.A.M. Saleh, *J. Lightwave Tech.* **11**(1), 42 (1993)

124. K. Feher, *Digital communications: Microwave applications.* (Prentice-Hall, Englewood Cliffs, NJ, 1981), pp. 71–106

125. W. Muys, J.C. Van der Plaats, F.W. Willems, H.J. Van Dijk, J.S. Leone, A.M.J. Koonen, *IEEE Photon. Tech. Lett.* **7**(6), 691 (1995)
126. J. Park, A.F. ELrefaie, K.Y. Lau, *IEEE Photon. Tech. Lett.* **9**(2), 256 (1997)
127. D. Wake, C.R. Lima, P.A. Davies, *IEEE Photon. Tech. Lett.* **8**(4), 578 (1996)
128. K. Yonenaga, N. Takachio, *IEEE Photon. Tech. Lett.* **5**(8), 949 (1993)
129. J. Park, W.V. Sorin, K.Y. Lau, *Electron. Lett.* **33**(6), 512 (1997)
130. U. Gliese, S.N. Nielsen, T.N. Nielsen, and frequency due to chromatic dispersion in fibre-optic microwave and millimeter-wave links," *1996 IEEE MTT-S Int. Microw. Symposium Digest (Cat. No.96CH35915)*, vol.3, pp. 1547-50, 1996, NY, USA
131. H. Schmuck, R. Heidemann, R. Hofstetter, *Electron. Lett.* **30**(1), 59 (1994)
132. A.A.M. Saleh, *Electron. Lett.* **25**(12), 776 (1989)
133. K. Kato, S. Hata, K. Kawano, J. Yoshida, A. Kozen, *IEEE J. Quantum Electron.* **28**(12), 2728 (1992)
134. M.P. Vecchi, *IEEE Commun. Mag.* **33**(8), 24 (1995)
135. W. Pugh, G. Boyer, *IEEE Commun. Mag.* **33**(8), 34 (1995)
136. C. Carroll, *IEEE Commun. Mag.* **33**(8), 48 (1995)
137. J. Park, M.S. Shakouri, K.Y. Lau, *Electron. Lett.* **31**(13), 1085 (1995)
138. W.I. Way, *IEICE Trans. Commun.* **E76-B**(9), 1091 (1993)
139. D.M. Fye, "Design of fiber optic antenna remoting links for cellular radio applications," *40th IEEE Veh. Technol. Cont*, pp. 622–625, Orlando, FL, May 1990
140. T.S. Chu, M.J. Gans, *IEEE Trans. Veh. Techn.* **40**(3), 599 (1991)
141. M. Shibutani, T. Kanai, W. Domom, K. Emura, J. Namiki, *IEEE J. Select. Areas Commun.* **11**(7), 1118 (1993)
142. M L.J. Greenstein, N. Amitay, T.S. Chu, L.J. Cimini, G.J. Foschini, M.J. Gans, I. Chih-Lin, A.J. Rustako, R.A. Valenzuela, G. Vannucci, *IEEE Commun. Mag.*, 76 (1992)
143. D. Parsons, *The Mobile Radio Propagation Channel.* (Halsted, New York, 1992)
144. A.J. Rustako Jr., N. Amitay, G.J. Owens, R.S. Roman, *IEEE Trans. Veh. Techn.* **40**(1), 203 (1991)
145. K. Petermann, E. Weidel, *IEEE J. Quantum Electron.* **17**, 1251 (1981)
146. B. Moslehi, *J. Lightwave Techn.* **4**, 1704 (1986)
147. J. Armstrong, *J. Opt. Soc. Am.* **56**, 1024 (1966)
148. M. Tur et al., "Spectral structure of phase-induced intensity noise in recirculating delay lines," in Proc Soc Photo-Optical Instrum Engr, April 4–8, 1983.
149. A. Yariv, Private discussions. Additional references are: Judy, A., "Intensity noise from fiber Rayleigh backscatter and mechanical splices," *Proc ECOC 1989*, paper TuP-11, Gimlet et al., "Observation of equivalent Rayleigh backscattering mirrors in lightwave systems with optical amplifiers," Photon. Technol. Lett., March 1990
150. J.L. Gimlett, N.K. Cheung, *J. Lightwave Tech.* **7**, 888 (1989)
151. A. Arie, M. Tur, *J. Lightwave Tech.* **8**, 1 (1990)
152. K. Vahala, A. Yariv, *IEEE J. Quantum Electron.* **QE-19**, 1102 (1983)
153. H.E. Rowe, *Signals and noise in communication systems.* (Van Nostrand, New York, 1965)
154. W. Harth, *Electron. Lett.* **9**, 532 (1973)
155. Gradshteyn, Ryznik, *Table of integrals, series and products.* (Academic, New York, 1980)

156. R. Epworth, *Laser Focus Mag.*, 109 (1981)

157. T. Wood, L. Ewell, *J. Lightwave Tech.* **4**, 391 (1986)

158. R.S.J. Bates, "Multimode waveguide computer data links with self pulsating laser diodes," *Proc. of Int. Topical Meeting on Optical Computing*, pp. 89–90, April 1990, (Kobe, Japan)

159. G.A. Wilson, R.K. DeFreeze, H.G. Winful, *IEEE J. Quantum Electron.* **27**, 1696 (1991)

160. U. Gliese, T.N. Nielsen, M. Bruun, E.L. Christensen, K.E. Stubkjer, S. Lindgren, B. Broberg, *IEEE Photon. Tech. Lett.* **4**, 936 (1992)

161. C.E. Zah, R. Bhat, S.G. Menocal, F. Favire, P.S.D. Lin, A.S. Gozdz, N.C. Andreadakis, B. Pathak, M.A. Koza, T.P. Lee, *Electron. Lett.* **27**, 1628 (1991)

162. K.Y. Lau, J.B. Georges, *Appl. Phys. Lett.* **63**(11), 1459 (1993)

163. R. Nagarajan, S. Levy, A. Mar, J.E. Bowers, *IEEE Photon. Tech. Lett.* **5**(1), 4 (1993)

164. O. Solgaard, J. Park, J.B. Georges, P.K. Pepeljugoski, K.Y. Lau, *Photon. Tech. Lett.* **5**(5), (1993)

165. S.D. Offsey, L.F. Lester, W.J. Schaff, L.F. Eastman, *Appl. Phys. Lett.* **58**(21), 2336 (1991)

166. H. Taub, L. Schilling, *Principles of Communication Systems*, 2nd edn. (McGraw Hill, New York, 1986), pp. 445–456

167. K. Simon, *Technical Handbook for CATV Systems*, 3rd edn. (General Instrument, 1996)

168. Some notes on composite second and third order intermodulation distortions, *Matrix Technical Notes MTN-108*, http://www.matrixtest.com/Literat/MTN108.htm, retrieved: 2008/2/15 12:17AM

169. Satcom and microwave fiber optics, http://www.emcore.com/product/fiber/satcom.php, retrieved: 2008/2/15 12:23AM

170. Model 2804/2805 CATV transmitter, http://www.emcore.com/assets/fiber/2804-2805slick.pdf, retrieved: 2008/2/15 12:24AM

171. New Focus : Products : Detectors : High-Speed Detectors and Receivers, http://www.newfocus.com/products/?navId=3&theView=listModelGroups&productLineId=3&productGroupId=135, retrieved: 2008/3/26, 9:37PM

172. T.E. Darcie, B.L. Kasper, J.R. Talman, C.A. Burrus, *J. Lightwave Tech.* **LT-5**(8), 1103 (1987)

173. R.C. Alferness, *IEEE Trans. Microw. Theory Tech.* **MTT-30**(8), 1121 (1982)

174. R.G. Walker, *IEEE J. Quantum Electron.* **27**(3), 654 (1991)

175. Y. Akage, K. Kawano, R. Iga, H. Ogamoto, Y. Miamoto, H. Takeuchi, *Electron. Lett.* **37**(5), 799 (2001)

176. L.N. Kurbatov, S.S. Shakhidzhanov, L.V. Bystrova, V.V. Krapukhin, S.J. Kolonenkov, *Soviet Phys. Semiconduct.* **4**, 1739 (1971)

177. T.P. Lee, C.A. Burrus, B.I. Miller, *IEEE J. Quantum Electron.* **9**, 829 (1973)

178. Amann, M. C., and Boeck, J., *Electron. Lett.* **15** 41 (1979)

179. M.C. Amann, J. Boeck, W. Harth, *Int. J. Electron.* **45**, 635 (1978)

180. M.C. Amann, A. Kuschmider, J. Boeck, *Electron. Lett.* **16**, 58 (1980)

181. W. Harth, M.C. Amann, *Electron. Lett.* **13**, 291 (1977)

182. M.C. Amann, J. Boeck, *AEU* **33**, 64 (1979)

183. K. Petermann, *IEEE J. Quantum Electron.* **15**, 566 (1979)

184. A.S. Daryoush, E. Ackerman, R. Saedi, R. Kunath, K. Shalkhauser, *IEEE Trans. Microwave Theory Tech.* **38**(5), 510 (1990)

185. E. Ackerman, D. Kasemset, S. Wanuga, R. Boudreau, J. Schlafer, R. Lauer, *IEEE Photon. Tech. Lett.* **3**(2), 185 (1991)
186. W. Ng, A.A. Waltson, G.L. Tangonan, J.J. Lee, I.L. Newberg, N. Bernstein, *J. Lightwave Tech.* **9**(9), 1124 (1991)
187. R. Adler, king phenomena in oscillators, *Proc. IRE*, 351 (1946)
188. J.B. Georges, K.Y. Lau, *IEEE Photon. Tech. Lett.* **4**(6), 662 (1992)
189. J.B. Georges, K.Y. Lau, *IEEE Photon. Tech. Lett.* **5**(2), 242 (1993)
190. X. Wang, G. Li, C.S. Ih, *J. Lightwave Tech.* **11**(2), 309 (1993)
191. D.B. Huff, J.P. Anthes, *IEEE Trans. Microwave Theory Tech.* **38**(5), 571 (1990)